Electromagnetism for Engineers

Andrew J. Flewitt
Department of Engineering and Sidney Sussex College
University of Cambridge
UK

This edition first published 2023
© 2023 John Wiley & Sons Ltd

The right of Andrew J. Flewitt to be identified as the author of this work has been asserted in accordance with law.

Registered Offices
John Wiley & Sons, Inc., 111 River Street, Hoboken, NJ 07030, USA
John Wiley & Sons Ltd, The Atrium, Southern Gate, Chichester, West Sussex, PO19 8SQ, UK

For details of our global editorial offices, customer services, and more information about Wiley products visit us at www.wiley.com.

Wiley also publishes its books in a variety of electronic formats and by print-on-demand. Some content that appears in standard print versions of this book may not be available in other formats.

Library of Congress Cataloging-in-Publication Data is Applied for:

Hardback ISBN: 9781119406167

Cover Design: Wiley
Cover Image: © Alexandr Gnezdilov Light Painting/Getty Images

Set in 9.5/12.5pt STIXTwoText by Straive, Chennai, India

For Hannah, Miriam, Tom and Rose

Contents

Preface

There are key evolutions in human technology that have had an impact upon everyday life. History recognizes these in terms such as the 'iron age' or the 'industrial revolution'. It could be argued that electromagnetism lies behind many of the really transformative technological changes since the early nineteenth century, when electromagnetism emerged as a science from what otherwise appeared to be a set of apparently magical phenomena. A list of key technologies that are enabled by electromagnetism should include the electric motor, the power station, radio transmission, the transistor and optical fibres. We live in an 'electromagnetic age'.

Electromagnetism is not only an important subject from a technological perspective, but also one of the most beautiful areas of physics. The Maxwell equations are very elegant mathematical expressions of basic properties of electric and magnetic fields, and, particularly in differential (vector calculus) form, they can be simply manipulated to explain a variety of physical electromagnetic phenomena.

The challenge that faces those trying to either teach or learn about electromagnetism is that it can be a rather abstract subject – an electron is not visible! This perception can create a mental barrier to understanding. However, many of the physical phenomena that flow from the fundamental principles are very familiar, from leaving a building to get a better mobile phone signal to knowing not to put your fingers between two magnets in case they snap together.

In this book, I have tried to take advantage of the power of mathematics to enable quantitative explanations of the physical phenomena, while also holding on to a physical reality, and I have included examples of real engineering that flows from these phenomena. My hope is that you will appreciate the wonder of electromagnetism both in the physics and especially in how we have been able to engineer this electromagnetic age.

Cambridge, UK   *Andrew J. Flewitt*
April 2022

Acknowledgements

This book has only been possible as a result of the support that I have received from both family and colleagues throughout the process of writing.

My wife Hannah has been wonderful throughout. She not only has been so understanding of the challenges of the writing process, but also went well outside her academic comfort zone to proof-read the whole book too! And to all my children who have been so enthusiastic of the idea of me writing a 'real book'.

Bill Milne gave me the confidence to embark on writing this book at the beginning of the whole process and kindly read the final version as well. Peter Davidson has also encouraged me of the value in writing this book as the few years that it has taken me have gone by.

My thanks too to all at Wiley for their patience and dedication in preparing this text.

Finally, I should thank all those who lectured me when I was reading physics as an undergraduate at Birmingham University. As I sat in the lecture theatres of the building named after John Henry Poynting, they brought electromagnetism to life and inspired my enjoyment of electromagnetism as a subject, which I hope I have managed to convey in some small measure in this book.

About the Author

Professor Andrew J. Flewitt read for a BSc in physics at the University of Birmingham before moving to the Engineering Department at Cambridge University where he gained his PhD. He was appointed a Fellow at Sidney Sussex College in 1999 and to a lectureship in the Engineering Department at Cambridge University in 2002, and has been teaching electromagnetism to students throughout this time, both in undergraduate college supervisions and in departmental lectures. His research group works on developing novel thin-film materials for microelectromechanical systems and large-area electronic devices. He was appointed Professor of Electronic Engineering in 2015 and Head of the Electrical Engineering Division in 2018.

Symbols

A	vector magnetic potential
A	area
dA	vector element of surface area
B	magnetic field as a vector quantity
B	magnitude of magnetic flux density
C	capacitance
c	wave velocity
D	electric flux density as a vector quantity
D	magnitude of electric flux density
d_{ij}	piezoelectric coefficients
E	electric field as a vector quantity
E	magnitude of electric field
e	magnitude of electronic charge
F	force as a vector quantity
F	magnitude of force
f	frequency
G	conductance
g	geometry factor
H	magnetic field as a vector quantity
H	magnitude of magnetic field
I	current
i	unit vector in the Cartesian x-direction
j	unit vector in the Cartesian y-direction
J	current density
I_1	line current
I_{ph}	phase current
j	$\sqrt{-1}$
k	unit vector in the Cartesian z-direction
k_B	Boltzmann constant
L	angular momentum
L	inductance
M	magnetization

$\mathbf{m}$	magnetic dipole moment
m_e	free mass of an electron
$\mathbf{N}$	Poynting vector
$\mathbf{P}$	polarization
P	real power
$\mathbf{p}$	dipole moment
Q	charge
Q	reactive power
q	charge
ρ	resistivity
$\mathcal{R}$	reluctance
R	resistance
$\mathbf{r}$	position vector
$\mathbf{dr}$	vector element of length
S	apparent power
T	temperature
T	time period
U	energy
V	potential difference
V_1	line voltage
V_{ph}	phase voltage
v	volume
Z	impedance
Z_g	waveguide impedance
Z_0	characteristic impedance
α	attenuation coefficient
β	propagation constant
β_g	propagation constant in a waveguide
γ	propagation constant
δ	skin depth
ε_0	permittivity of free space
ε_r	relative permittivity
η	intrinsic impedance
λ	wavelength
λ_g	wavelength in a waveguide
μ_0	permeability of free space
μ_r	relative permeability
ρ	volume charge density
ρ_L	reflection coefficient
ρ_T	transmission coefficient
σ	surface charge density
σ_P	surface polarization charge
Φ	total flux through a loop
$\cos\phi$	power factor
χ_B	magnetic susceptibility
χ_E	electric susceptibility
ω	angular frequency

About the Companion Website

This book is accompanied by a companion website:

www.wiley.com/go/flewitt/electromagnetism

From the website you can find the following online materials:
- Questions and Answers
- Videos

Part I

Fundamentals of Electricity and Magnetism

1

Charge and Electric Fields

1.1 Charge as a Fundamental Property of Matter

All matter is comprised of fundamental sub-atomic particles which themselves have basic properties, such as mass, charge and spin. The sub-atomic particle which electrical engineers are probably most concerned with is the electron, which was discovered by the Nobel Prize winning physicist J.J. Thomson in Cambridge in 1897 (Thomson 1897). Among its fundamental properties, electrons have a mass of 9.109×10^{-31} kg and a charge of -1.602×10^{-19} C, the magnitude of which we call e. Of these two properties, we are probably more familiar with the concept of mass as the world around us is dominated by gravitational forces at the macroscopic scale. We understand that two particles which have non-zero mass both experience a force between them. We rationalize this by saying that a particle with mass produces a gravitational 'field' which extends spatially away from the particle. Another particle with a non-zero mass inside this field then experiences a force.

As engineers, we do not tend to worry about the exact mechanism by which this force is being exerted over some distance, leaving that important consideration to physicists. Instead, we simply apply the equations for force between masses, such as that exerted by the earth on all structures in civil engineering. Therefore, we should be content to accept the less familiar concept of charge on the same basis: a particle with a non-zero charge produces an *electric field* which extends spatially away from the particle, and another particle with a non-zero charge inside this field then experiences a force. We use the symbol Q or q to denote charge, and the SI unit of charge is the *coulomb* (C).

1.2 Electric Field and Flux

Fields are widely used in physics to describe regions of space in which an object with a particular property experiences a force. Therefore, an electric field is a region of space in which an object with charge q experiences a force $\mathbf{F}$. As force is a vector quantity, having both a magnitude and direction, electric field must also be a vector quantity $\mathbf{E}$, so that

$$\mathbf{F} = \mathbf{E}q \tag{1.1}$$

From this equation, it can be seen that the unit of electric field is N C^{-1}, although, as we will see in Section 1.3, the more common unit is V m^{-1}.

Electromagnetism for Engineers, First Edition. Andrew J. Flewitt.
© 2023 John Wiley & Sons Ltd. Published 2023 by John Wiley & Sons Ltd.
Companion website: www.wiley.com/go/flewitt/electromagnetism

In everyday life we experience objects with both positive and negative charge, because while electrons have a charge of $-e$, protons, which are one of the sub-atomic constituents of the nucleus of atoms, have a positive charge of e. Therefore, the force acting on a positive charge at a particular point in an electric field will act in the opposite direction to the force on a negative charge at the same point. This is the reason why like charges repel each other whereas opposite charges attract. This is in contrast to mass, which is positive for all matter, and therefore the force between masses is always attractive.

To assist us in visualizing fields, we use the concept of *flux*. We imagine that the electric field is composed of lines of flux whose direction at a given point in space is the direction of the electric field at that point and whose number density per unit area relates to the magnitude (or intensity) of the electric field.

We know that charge produces electric fields, and therefore lines of electric flux begin on positive charges and end on negative charges. As the total sum of all charge in the universe is zero, it must be the case that every line of flux that begins on a positive charge must have a balancing negative charge somewhere to end on. We can now visualize the electric field around a small point charge $+q$ in free space (a vacuum) in Figure 1.1. If we assume that the balancing charge of $-q$ is uniformly distributed an infinite distance away, then lines of electric flux will radiate uniformly away from the point charge. This will cause the field to decrease with increasing radial distance r from the point charge as $1/r^2$, just as we find for gravitational fields around mass as well. This is the basis behind the *Coulomb law* for the magnitude of the force F that acts on a charge q_2 in an electric field of magnitude E_1 produced by another point charge q_1:

$$F = E_1 q_2 = \left(\frac{q_1}{4\pi\varepsilon_0 r^2} \right) q_2 \tag{1.2}$$

where r is the distance between the charges.

In Eq. (1.2) we have had to introduce a new quantity ε_0, which is the *permittivity of free space*. It is a fundamental constant with the value 8.854×10^{-12} F m^{-1}, and it is required to yield a result for the force between two charges that is correct in SI units.

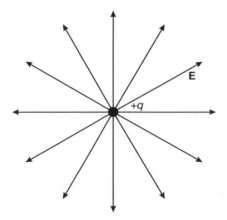

Figure 1.1 Lines of electric flux around a point charge $+q$ in free space assuming that the balancing charge is uniformly distributed an infinite distance away. The direction and area density of the flux lines are a measure of the electric field **E** at any point.

1.3   Electrical Potential

Let us imagine that we have two point charges of equal magnitude but opposite sign, $+q$ and $-q$, which initially exist at the same point in free space, so that there is no electric field around the charges. If we slowly move the charge $-q$ away from the $+q$ charge so that the distance r between them is increasing, then an electric field distribution will be created around and between the charges as shown in Figure 1.2. As there will be a force acting on the charge that is being moved, given by Eq. (1.2), there must be some change in energy – work is being done against the force that is attracting the charges together. In this case, mechanical energy is being converted into an electrical potential energy, which could be converted back into mechanical energy again by allowing the two charges to accelerate back towards each other once more. It is a key concept that whenever a field (whether electric, magnetic or gravitational) occupies a volume of space, then some potential energy has been stored.

We can use basic mechanics to relate electric field to potential energy. If we have a charge q, in an electric field of magnitude E, which is moved by a small distance δx in the direction of the electric field, then there will be a change in potential energy of the charge given by

$$\delta W = -F\delta x = -Eq\delta x \tag{1.3}$$

where F is the magnitude of the force acting on the charge due to the electric field. The change in potential energy is negative as the force acting on the charge is in the same direction as the movement. We define a new quantity, the *potential difference*, which is the change in energy per unit charge between two points in space. The potential difference is given the symbol V and has units of volts. Therefore, the small potential difference δV between the two points separated by the distance δx over which we have moved our charge q is

$$\delta V = \frac{\delta W}{q} \tag{1.4}$$

Equating δW in Eq. (1.3) and (1.4) gives

$$q\delta V = -Eq\delta x \tag{1.5}$$

and therefore, by basic calculus, we have the result that

$$E = -\frac{dV}{dx} \tag{1.6}$$

In other words, the electric field is the negative of the potential gradient. For readers who are familiar with vector calculus, we can rewrite this in three dimensions as

$$\mathbf{E} = -\nabla V \tag{1.7}$$

It should be noted that we can only ever talk about a potential difference *between two points*, for example the potential at a point a with respect to a point b which we could denote V_{ab}. The direction is significant, as $V_{ba} = -V_{ab}$. We often use arrows to denote a potential difference where we are considering the potential at the tip of the arrow with respect to the tail. If we know the electric field distribution between the two

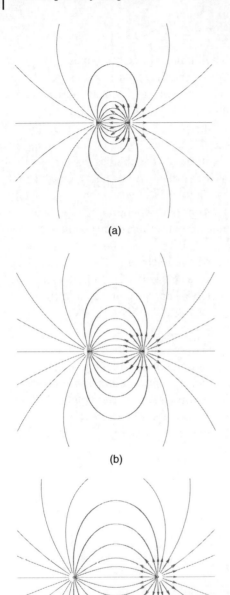

Figure 1.2 The electric field distribution around two point charges $+q$ (right) and $-q$ (left) separated by increasing distance from (a) to (c).

(a)

(b)

(c)

points, then we can evaluate this. A simple integration of the two sides of Eq. (1.6) would suggest that

$$V_{ab} = -\int_b^a E \, dx \tag{1.8}$$

However, this is a slight simplification as we know that the electric field is actually a vector quantity, and it is therefore only the component of a small movement dx parallel to the direction of the electric field that will lead to a change in potential. If we use vectors to express both the electric field $\mathbf{E}$ and the small movement $\mathbf{dx}$, then the scalar (dot) product of the two yields exactly this result. Therefore, Eq. (1.8) is more generally expressed as

$$V_{ab} = -\int_b^a \mathbf{E} \cdot \mathbf{dx} \tag{1.9}$$

In practice, Eq. (1.8) is more commonly used in simple calculations. This is achieved by ensuring that a path between two points is chosen so that $\mathbf{E}$ and $\mathbf{dx}$ are always either parallel or perpendicular to each other so that the scalar product is either a simple product or zero at any point.

This picture is very similar to the situation for gravity, where we can define a gravitational potential difference as being the difference in gravitational potential energy per unit mass between two points. In practice, we often take the surface of the earth as a reference point from which to measure gravitational potential as a function of height h above the surface of the earth which, assuming the gravitational constant g to be uniform, is simply gh. Likewise, it is helpful to take a constant reference point from which to measure electric potential difference. The earth again provides a good practical reference point, as it is so large that small changes in the charge on the earth have no significant impact upon its electrical potential energy. For other situations, such as the point charge in free space shown in Figure 1.1, the earth is simply not present, and so we have to choose an alternative reference from which to measure potential difference. In this example, an infinite distance away from the charge ($r = \infty$) provides a good reference point. We can therefore use Eq. (1.8) to calculate an expression for the potential difference with respect to this reference $V(r)$ around the point charge in free space from the expression for the electric field around the charge in Eq. (1.2):

$$V(r) = -\int_\infty^r \frac{q_1}{4\pi\varepsilon_0 r^2} dr = \left[\frac{q_1}{4\pi\varepsilon_0 r}\right]_\infty^r = \frac{q_1}{4\pi\varepsilon_0 r} \tag{1.10}$$

Note that we have been able to use the scalar Eq. (1.8) rather than the vector Eq. (1.9) by choosing a radial path which is parallel to the radial electric field. To help visualize potentials, we often plot lines of equipotential, rather as we plot contour lines on a map of constant height and also therefore constant gravitational potential, and this is shown for the point charge in Figure 1.3. As electric field points down a potential gradient according to Eqs. (1.6) and (1.7), lines of electric flux always cut perpendicularly through lines of equipotential.

It is clear from Eq. (1.10) that the potential difference with respect to $r = \infty$ at any point in free space around the point charge is uniquely defined. This is always true as we are dealing with a linear system. Therefore, it does not matter what path is actually chosen over which to evaluate the potential difference between two points using the integral in Eq. (1.9) – the result will always be the same. It is for the same reason that if you walk up a hill, the change in your gravitational potential energy will not depend on which path you

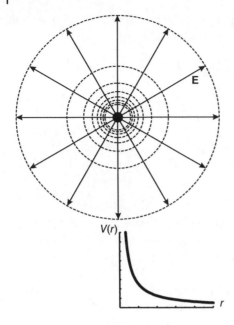

Figure 1.3 Lines of electric flux around a point charge $+q$ in free space with dashed lines of equipotential calculated with respect to $r = \infty$, with the graph of how potential varies as a function of distance r from the charge shown underneath on the same length scale.

take. Likewise, if we choose a closed path that ends back at its starting point, then there will be no potential difference between the start and end of the loop, which can be expressed mathematically from Eq. (1.9) as

$$\oint_C \mathbf{E} \cdot \mathbf{dx} = 0 \tag{1.11}$$

where $\oint_C$ indicates the integral round a closed loop.

1.4 The Gauss Law of Electrostatics in Free Space

Our starting point for the discussion about charge as a fundamental property of matter in Section 1.1 was that charge produces an electric field. In the following discussion we then refined this to say that lines of electric flux begin on positive charges and end on negative charges. The *Gauss law of electrostatics* takes this basic statement about the origin and nature of electric fields and expresses it in a rather elegant mathematical formulation, namely,

$$\oint_S \mathbf{E} \cdot \mathbf{dA} = \frac{q}{\varepsilon_0} \tag{1.12}$$

where the left-hand side is an integral over a closed *Gaussian surface* and q is the net charge enclosed by the surface. We will now consider the mathematical basis for this equation.

Let us imagine that we have an arbitrary surface in three dimensions, such as that in Figure 1.4. We could work out its surface area by laying it out onto a flat surface and measuring its area, but calculus and vector algebra allow us to take a more elegant approach. We could imagine dividing up the surface into lots of small elements of the surface, each of which approximates to a flat parallelogram, as shown in Figure 1.4. We define two vectors

Figure 1.4 An arbitrary surface divided up into small parallelograms. For one parallelogram, the side vectors have been shown as small arrows and the resulting vector element of surface area **dA** is shown.

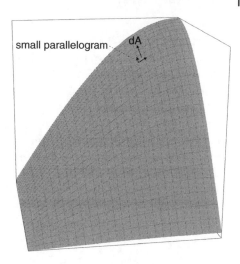

which point along adjacent edges of the small parallelogram; if we take the cross product of these two vectors then the result is a third vector which points perpendicularly away from the plane of the parallelogram and whose magnitude is equal to the area of the parallelogram. We call this vector **dA**. If we were to integrate the magnitude of all of the **dA** vectors over the whole surface, then we would have the whole surface area,

$$\int_{surface} |\mathbf{dA}| = \text{total surface area} \tag{1.13}$$

Let us now imagine that an electric field is passing through the surface. Taking the dot product of two vectors, such as **a** and **b**, gives the component of **a** in the direction of **b** multiplied by the magnitude of **b**, or, expressed mathematically,

$$\mathbf{a} \cdot \mathbf{b} = |\mathbf{b}|(|\mathbf{a}| \cos \theta) \tag{1.14}$$

where θ is the angle between **a** and **b**. Therefore, if we take the dot produce of **E** with any one of the small **dA** vectors, the result will be the component of **E** perpendicular to the surface multiplied by the area of the small surface element. Therefore, as |**E**| is the number of flux lines per unit area, **E** · **dA** is the number of flux lines passing perpendicularly through the small element of surface. We are taking a very similar approach to that used when calculating the force exerted by a ladder that is resting against a wall. For this mechanics problem we would resolve the force to calculate the component of force acting perpendicular to the surface of the wall. Here we are just resolving the electric field to take the component perpendicular to the surface.

We can therefore calculate the total number of electric flux lines passing perpendicularly through any surface by simply integrating **E** · **dA** over the surface,

$$\int_{surface} \mathbf{E} \cdot \mathbf{dA} = \text{total flux through surface} \tag{1.15}$$

Therefore, if the surface is closed (such as a box, sphere, or cylinder) then this integral would give the total flux passing out of the surface. We should note that the mathematical convention is that vector elements of surface area, **dA**, always point out of closed surfaces.

We should now be able to intuitively understand the meaning of the Gauss law of electrostatics as expressed in Eq. (1.12). We can call any closed surface that we can imagine in space a Gaussian surface. By taking the integral of $E \cdot dA$ over that surface, we are effectively counting flux lines. A line leaving the volume enclosed by the surface counts positively, while one entering counts negatively. As lines of electric field begin and end on charges, if there is no net charge enclosed within the Gaussian surface then the result of the integral must be zero. As many lines of flux leave through the Gaussian surface as enter. The result can only be non-zero if a line of flux has either begun or ended on a charge that is somewhere in the volume enclosed by the Gaussian surface. This is exactly what Eq. (1.12) states: that $\oint_S E \cdot dA$ is proportional to the net charge q enclosed by the surface. We only now need $1/\varepsilon_0$ as a constant of proportionality to ensure that the correct numerical values are produced.

1.5 Application of the Gauss Law

The Gauss law of electrostatics allows us to calculate the electric field produced by distributions of charge in space, and consequently the potential using Eqs. (1.6) or (1.7). The presence of the surface integral in Eq. (1.12) can make the Gauss law appear rather intimidating in the first instance. However, in practice we can often calculate the electric field produced by quite complex charge distributions through careful choice of the Gaussian surface. This may be achieved by ensuring that the electric field and vector elements of surface area are either parallel or perpendicular to each other, normally by reflecting the geometry of the charge distribution in that of the Gaussian surface.

For example, we can very easily derive the expression for the electric field around a point charge that is implicit in the Coulomb law (Eq. (1.2)). Intuitively, we would expect a point charge q to produce a spherically symmetric electric field, as shown in Figure 1.3. Therefore, we should choose a Gaussian surface that is a sphere of radius r centred on the charge, where the Gauss law will allow us to evaluate the electric field at that radius. As both the electric field E and the surface vectors dA both point radially outwards, the dot product of E and dA becomes just a simple product. Also, as can be clearly seen in Figure 1.3, the magnitude of the electric field $E(r)$ is a constant at a given radius. Therefore, for this scenario using Eq. (1.13) we have

$$\oint_S E \cdot dA = E(r) \oint_S |dA| = E(r) \times \text{area} \tag{1.16}$$

As the area of the sphere is $4\pi r^2$, Eq. (1.12) reduces simply to

$$E(r) \cdot 4\pi r^2 = \frac{q}{\varepsilon_0} \tag{1.17}$$

which then rearranges to the familiar expression

$$E(r) = \frac{q}{4\pi \varepsilon_0 r^2} \tag{1.18}$$

We can also handle more complex charge distributions, such as an infinitely long line of charge with a charge per unit length of σ. In this case, the system has cylindrical symmetry,

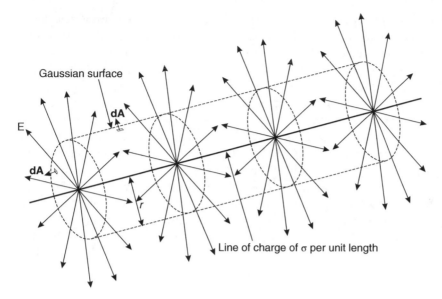

Gaussian surface

E

dA

dA

r

Line of charge of σ per unit length

Figure 1.5 Lines of electric flux around a line of charge of $+\sigma$ per unit length in free space. A cylindrical Gaussian surface of length l and radius r is shown using a dashed line. Two elements of surface area **dA** are shown: one on an end of the cylinder pointing axially outwards and one on the curved surface pointing radially outwards.

and so we should choose a Gaussian surface which is a cylinder axially centred on the line of charge, with some arbitrary length l and a radius r, as shown in Figure 1.5. In this case, the vector elements of surface area on the end caps of the cylinder will point in the direction of the axial line of symmetry of the system, whereas the electric field will point radially outwards. Hence, $\mathbf{E} \cdot \mathbf{dA} = 0$, so these do not need to be considered further. Over the curved surface of the cylinder, $\mathbf{E}$ and the surface vectors $\mathbf{dA}$ both point radially outwards, and so we can use Eq. (1.16) again. The area of the curved surface of the cylinder is $2\pi rl$, and therefore the Gauss law reduces to

$$E(r) \cdot 2\pi rl = \frac{\sigma l}{\varepsilon_0} \tag{1.19}$$

where σl is the charge enclosed within the cylinder. This rearranges to

$$E(r) = \frac{\sigma}{2\pi\varepsilon_0 r} \tag{1.20}$$

Therefore, although Eq. (1.12) may appear rather complex at first sight, involving surface integrals of vector quantities, in many cases the actual application of the Gauss law only requires scalar multiplication of fields and areas.

1.6 Principle of Superposition

We have seen that the Gauss law of electrostatics can be used to calculate the electric field around geometrically 'simple' distributions of charge. We can then use this as a basis for calculating the field around more complex distributions using the *principle of superposition*.

The principle of superposition in its most general form as it applies across diverse branches of physics states that if there is a linear relationship between some stimulus and a response, then the total response due to many stimuli acting simultaneously is the same as the sum of the responses that each stimulus would have produced individually. In the case here, the stimulus is the charge which produces a response in the form of an electric field, and the Gauss law in Eq. (1.12) clearly shows that these are proportional to each other (in other words, they are linearly related). Therefore, we can break down a complex charge distribution into many small elements of charge, calculate the electric field distribution due to each of the elements and then simply add all the contributions to the electric field at each point in space to yield the total field at that point.

We could have used superposition to calculate the electric field around the line of charge shown in Figure 1.5 by dividing the line up into infinitesimally small elements of charge and then integrating the resulting fields at any point to determine the total field, but the symmetry of this charge distribution makes the approach using the Gauss law directly a significantly simpler means of deriving Eq. (1.20).

1.7 Electric Dipoles

Two charges of equal magnitude but opposite sign separated by a small distance are called a *dipole*. The electric field distribution around the dipole of two point charges $+q$ and $-q$ in Figure 1.2 has been calculated using the Principle of Superposition. Each of the two charges would independently produce their own electric field distribution, one pointing radially outwards and the other radially inwards, as shown in Figure 1.6. If we imagine the plane passing through all points that are equidistant between these two charges, then the components of the electric fields due to the two charges pointing parallel to this plane will cancel out when superposed, leaving only the perpendicular components which sum, as is clearly the case in Figures 1.2 and 1.6. If the two charges are separated by a distance d then, from Eqn. (1.18), the electric field at the equidistant point along the line between the two charges due to each of the two charges independently will be

$$E_+ = \frac{q}{4\pi\varepsilon_0(d/2)^2} \tag{1.21a}$$

$$E_- = \frac{-q}{4\pi\varepsilon_0(d/2)^2} \tag{1.21b}$$

To calculate the total electric field at this point, we would need to sum Eqs. (1.21a) and (1.21b). However, the two expressions have been calculated in different frames of reference: one pointing away from the positive charge and one pointing away from the negative charge. We have to use a common frame of reference when summing the two fields. In the frame of reference used for the positive charge, the electric field due to the negative charge at the equidistant point is $-E_-$, so the total field is

$$E = E_+ - E_- = \frac{2q}{\pi\varepsilon_0 d^2} \tag{1.22}$$

Therefore, the electric field at this equidistant point decreases as the distance between the dipole increases, and this can be seen in the density of the flux lines around this point

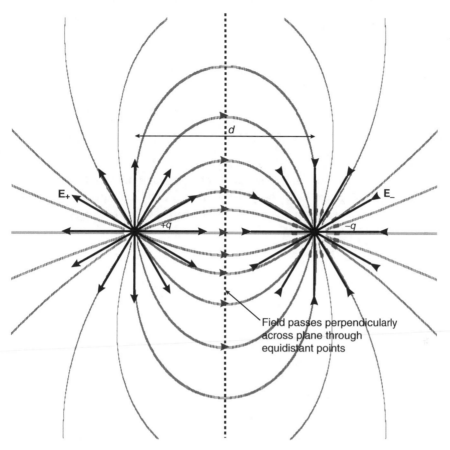

Figure 1.6 The electric fields **E**$_+$ and **E**$_-$ produced separately by each of the point charges in a dipole and the resultant total electric field calculated by superposition. The field passes perpendicularly through the plane of equidistance between the two charges.

shown in Figure 1.2, which are densest in Figure 1.2a where the dipole separation is smallest. At long distances away from the dipole (much greater than d) the two charges appear to be at the same point in space, and their electric fields will cancel out to leave no net field, and so the electric field is effectively concentrated in the region of space close to the two charges. This effect is also most pronounced in Figure 1.2a.

In practice, dipoles frequently appear in real situations as most objects have no net charge, and so the act of removing some charge by some distance from a particular object immediately creates a dipole between the removed charge and the original object. At the atomic scale, when an atom experiences an externally applied electric field, the electrons around the nucleus will be displaced from their equilibrium position, as shown in Figure 1.7. Just as we can think of a distributed mass acting from a 'centre of mass' point, so we can think of the electrons around the nucleus as acting from a 'centre of charge'. The action of the electric field is therefore to move this effective centre of charge by a vector displacement **d** from the positive charge, resulting in the formation of a dipole.

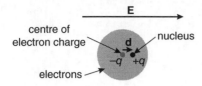

Figure 1.7 A simplistic diagram of the effect of an externally applied electric field **E** on a single atom, which results in an effective separation of the positive charge $+q$ on the nucleus and the negative charge $-q$ of the electrons by a displacement vector **d**, which results in a dipole moment $\mathbf{p} = q\mathbf{d}$.

A helpful quantity that is often used when performing calculations involving dipoles is the *dipole moment* **p**. It is a vector quantity that is simply defined as the product of the magnitude of the dipole charge q and the vector displacement of the positive charge from the negative charge **d**, that is,

$$\mathbf{p} = q\mathbf{d} \tag{1.23}$$

The dipole moment has units of C m.

Reference

Thomson, J.J. (1897). XL. Cathode rays. *Philosophical Magazine Series 5* 44 (269): 293–316.

2

Electric Fields in Materials

2.1 The Interaction of Electric Fields with Matter

In Chapter 1, we considered the consequences of the basic statement that charge produces electric fields and we saw that the Gauss law of electrostatics expresses this in a simple mathematical form. However, throughout this discussion we confined ourselves to considering the most basic situation that the charge and associated electric field exist in a vacuum. In practice, however, charges and fields exist in real matter, and the response of this matter to the electric field has a profound influence upon the field itself. We shall start by considering simple insulating materials, commonly called *dielectrics*, and a key electronic device which employs dielectrics – the *capacitor*. This leads on to a discussion on *ferroelectric* and *piezoelectric* materials. We will then move on to the other extreme situation of metals, which are highly conducting, before concluding with *semiconductors*, which underpin solid state electronic devices.

2.2 Dielectrics

Dielectrics are materials that have no free charges in their bulk. As a consequence they conduct the flow of current very poorly, and are commonly called insulators. Such materials are characterized by there being a large energy barrier preventing *valence electrons* (those in the outermost shell of electrons with the highest energy states around an atom) from escaping from localized *bonding states* into delocalized *antibonding states*. For example, in covalently bonded materials, there is usually such an energy barrier as the covalent bond is a deep potential well in which the bonding electrons are trapped, and so most plastic materials are dielectrics and are widely used as the insulators in flexible cables. Other examples of dielectrics include silicon dioxide (glass) which has historically been used as the insulator in silicon field effect transistors, such as would be found in microprocessors or memory devices. More recently silicon dioxide has been displaced in field effect transistors by other materials including oxides of metals, many of which are also dielectrics.

Although there is no free charge that can move on macroscopic scales in dielectrics, electrons are displaced locally within their potential wells in the presence of an electric field due to the force that acts on them, as shown in Figure 1.7. As a result, many local dipole moments **p** are created (see Section 1.7) when a dielectric is subjected to an externally

Electromagnetism for Engineers, First Edition. Andrew J. Flewitt.
© 2023 John Wiley & Sons Ltd. Published 2023 by John Wiley & Sons Ltd.
Companion website: www.wiley.com/go/flewitt/electromagnetism

applied electric field. If there are N such dipole moments created per unit volume, then we can define a *polarization* $\mathbf{P}$ of the material due to the applied field as the dipole moment per unit volume which will be given by

$$\mathbf{P} = N\mathbf{p} \qquad (2.1)$$

and this will have units of $\mathrm{C\,m^{-2}}$.

Figure 2.1a shows the example of a block of dielectric material in the presence of a uniform external electric field $\mathbf{E}_{ext}$. In the bulk of the dielectric, there is no change in the total charge present, and so the total negative charge will still exactly balance the total positive charge, leading to net neutrality. However, on the surface of the dielectric, this local charge displacement will lead to the formation of a *surface polarization charge*, which we will call σ_P per unit area.

As opposite charges attract, this surface polarization charge will cause the electric field inside the dielectric to be reduced compared to that outside. To understand this quantitatively, it is simplest to use superposition (see Section 1.6) by first considering the field induced by the surface charge within the dielectric, which is called the polarization

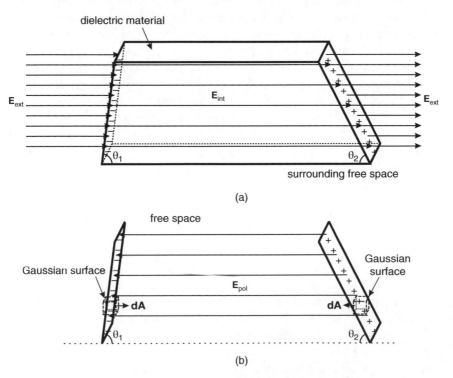

(a)

(b)

Figure 2.1 (a) A uniform external electric field $\mathbf{E}_{ext}$ is applied to a block of a dielectric material which will then have a different internal electric field $\mathbf{E}_{int}$ due to the formation of a surface polarization charge σ_P per unit area. The ends of the dielectric material are at angles of θ_1 and θ_2 from the electric field. (b) We now imagine removing the external field and the block of dielectric material but leave behind just the two layers of surface polarization charge. We also imagine two small cubic Gaussian surfaces, one crossing each of the layers of charge, which will allow us to calculate the polarization field $\mathbf{E}_{pol}$.

field $\mathbf{E}_{pol}$, as shown in Figure 2.1b. We can then simply add the applied external field $\mathbf{E}_{ext}$ to this to get the dielectric's actual internal electric field $\mathbf{E}_{int}$.

We can use the Gauss law of electrostatics to calculate $\mathbf{E}_{pol}$ by considering a small cubic Gaussian surface covering an area of each of the charged ends of the dielectric in Figure 2.1b. Each surface of the cube has an area $|\mathbf{dA}| = dA$. For the left-hand side, only one surface of the cube has a field passing through it of $\mathbf{E}_{pol}$. The charge enclosed by the Gaussian surface is the product of the surface polarization charge on that surface, which we will call $-\sigma_{P1}$ (noting that this is a negative charge) and the area of the charged surface enclosed, which from basic geometry will be $dA/\sin\theta_1$ due to the face of the dielectric being at an angle of θ_1 from the bottom face. Therefore, for the left-hand face, a simple application of the Gauss law of electrostatics (Eq. (1.12)) gives

$$\oint_S \mathbf{E} \cdot \mathbf{dA} = \frac{q}{\varepsilon_0}$$

$$-|\mathbf{E}_{pol}|dA = \frac{-\sigma_{P1}dA}{\varepsilon_0 \sin\theta_1} \tag{2.2}$$

We should note that as $\mathbf{E}_{pol}$ is pointing to the left and $\mathbf{dA}$ is pointing to the right, the scalar (dot) product of the two is $-|\mathbf{E}_{pol}|dA$. Eq. (2.2) then simply reduces to

$$|\mathbf{E}_{pol}| = \frac{\sigma_{P1}}{\varepsilon_0 \sin\theta_1} \tag{2.3}$$

A similar application of the Gauss law of electrostatics to the right-hand side gives

$$|\mathbf{E}_{pol}| = \frac{\sigma_{P2}}{\varepsilon_0 \sin\theta_2} \tag{2.4}$$

Equations (2.3) and (2.4) can be equated as $\mathbf{E}_{pol}$ is uniform in the region of space that was occupied by the dielectric material:

$$\frac{\sigma_{P1}}{\sin\theta_1} = \frac{\sigma_{P2}}{\sin\theta_2} \tag{2.5}$$

The two surfaces of the dielectric will have different surface polarization charges per unit area, but the total charge on the two surfaces are equal in magnitude. In fact, the quantity $-\sigma_P/\sin\theta_1$ in Eq. (2.5) is the surface charge per unit cross sectional area perpendicular to the applied electric field of the dielectric block, which in turn is just the polarization $\mathbf{P}$ (remember that polarization points from negative to positive charges by definition from Section 1.7). Therefore, substituting this back into Eq. (2.3) or (2.4) gives

$$\mathbf{E}_{pol} = -\frac{\mathbf{P}}{\varepsilon_0} \tag{2.6}$$

We should note that $\mathbf{E}_{pol}$ and $\mathbf{P}$ must point in opposite directions as electric fields point from positive to negative charges (see Section 1.2 for the definition), but the polarization points from negative to positive charges (see Section 1.7 for the definition).

We can now relate the actual internal electric field in the dielectric, the externally applied electric field and the polarization field. We know that the electric field inside the dielectric material is reduced by some factor compared with the external field, which is called the *relative permittivity* and is given the symbol ε_r, so that

$$\mathbf{E}_{int} = \frac{\mathbf{E}_{ext}}{\varepsilon_r} \tag{2.7}$$

Superposition (see Section 1.6) now allows us to determine the dielectric's internal electric field as it is simply the combination of the applied external electric field and the polarization field

$$\mathbf{E}_{int} = \mathbf{E}_{ext} + \mathbf{E}_{pol} \tag{2.8}$$

Therefore, we can substitute Eqs. (2.6) and (2.7) into Eq. (2.8) to eliminate $\mathbf{E}_{ext}$ to give an expression for how the internal electric field relates to the polarization:

$$\mathbf{E}_{int} = \varepsilon_r \mathbf{E}_{int} - \frac{\mathbf{P}}{\varepsilon_0}$$

$$\mathbf{P} = (\varepsilon_r - 1)\varepsilon_0 \mathbf{E}_{int} \tag{2.9}$$

The factor $\varepsilon_r - 1$ is often called the *electric susceptibility* of the material and is given the symbol χ_E.

Calculating the electric field in a system with a number of different dielectric materials is a challenge due to the presence of surface charges at every boundary. By way of an example, let us consider again the small Gaussian surface on the right-hand side of the block of dielectric material shown in Figure 2.1. The Gauss law of electrostatics (Eq. (1.12)) states that

$$\oint_S \mathbf{E} \cdot \mathbf{dA} = \frac{q}{\varepsilon_0} \tag{2.12}$$

Clearly for this Gaussian surface there is a charge on the surface of the dielectric caused by its polarization. We have already seen that the polarization $\mathbf{P}$ is just the surface charge per unit cross-sectional area perpendicular to the applied electric field, so for this Gaussian surface which is a cube with sides of area $|\mathbf{dA}|$, the surface charge enclosed, q_s, is just

$$q_s = -\oint_S \mathbf{P} \cdot \mathbf{dA} \tag{2.10}$$

If there is any other free charge q_f inside the Gaussian surface that is not associated with the surface polarization, then Eq. (1.12) becomes

$$\oint_S \varepsilon_0 \mathbf{E} \cdot \mathbf{dA} = q_f + q_s = q_f - \oint_S \mathbf{P} \cdot \mathbf{dA} \tag{2.11}$$

This rearranges to

$$\oint_S (\varepsilon_0 \mathbf{E} + \mathbf{P}) \cdot \mathbf{dA} = q_f \tag{2.12}$$

We therefore define a new quantity, the *electric flux density* $\mathbf{D}$, which is defined as

$$\mathbf{D} = \varepsilon_0 \mathbf{E} + \mathbf{P} \tag{2.13}$$

and which has units of C m^{-2}. Therefore, substituting this into Eq. (2.12) gives a new form of the Gauss law of electrostatics in terms of the electric flux density:

$$\oint_S \mathbf{D}.\mathbf{dA} = q_f \tag{2.14}$$

This equation allows **D** to be calculated for a distribution of free charges in the presence of dielectric materials without the need to calculate either the polarization or any surface charge densities. Furthermore, substituting Eq. (2.9) into Eq. (2.13) allows **P** to be eliminated to give

$$\mathbf{D} = \varepsilon_0 \varepsilon_r \mathbf{E} \tag{2.15}$$

Therefore, if the relative permittivity of the dielectric material occupying a particular point in space is known, and having calculated **D** at that point, evaluating **E** is trivial.

For example, let us consider the situation shown in Figure 2.2 where there is a dipole of two charges, $+q$ and $-q$, in free space which are separated by a distance d. A plate of a dielectric material with a relative permittivity ε_r is placed at the equidistant plane between the two charges. We wish to know the maximum electric field inside the dielectric plate, which will be at the point along the line between the two charges. A 'brute force' application of the Gauss law of electrostatics as in Eq. (1.12) would require us to calculate the surface polarization charge on the dielectric plate, which will vary with position over the surface. However, if we use the form of the Gauss law in Eq. (2.14), then the problem becomes much easier. We can calculate the electric flux density due to each of the two charges independently at the maximum field point and then superpose the two to get the total electric flux density. We

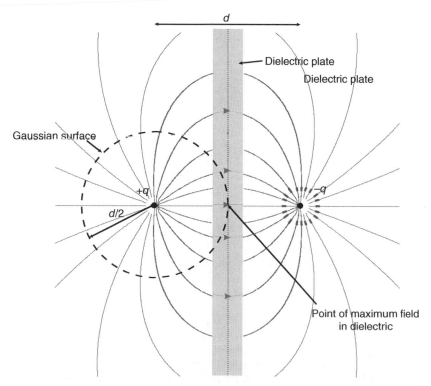

Figure 2.2 The electric field produced by a dipole of two charges in free space with a plate of dielectric material between the two.

use a spherical Gaussian surface centred on each charge with a radius $d/2$. For the positive charge, applying Eq. (2.14) gives

$$|\mathbf{D}_+|4\pi r^2 = +q$$

$$|\mathbf{D}_+| = \frac{q}{4\pi r^2} \tag{2.16}$$

The similar expression for the negative charge will be

$$|\mathbf{D}_-| = \frac{-q}{4\pi r^2} \tag{2.17}$$

In combining the two, we must remember that the direction of the electric flux density in each case points away from the charge, and therefore the two frames of reference point in opposite directions at the point of maximum electric field in the dielectric. If we determine the total electric flux density in the frame of reference used for the positive charge, then

$$|\mathbf{D}| = |\mathbf{D}_+| - |\mathbf{D}_-| = \frac{q}{2\pi r^2} \tag{2.18}$$

Using Eq. (2.15), the electric field at this point is simply

$$|\mathbf{E}| = \frac{q}{2\pi\varepsilon_0\varepsilon_r r^2} \tag{2.19}$$

Eq. (2.15) implies that there is a simple linear relationship between the electric field and electric flux density for all situations. However, in practice this is not actually the case. All dielectrics have a maximum *breakdown field* that they can withstand before significant conduction takes place, in a phenomenon known as *dielectric breakdown*. Atmospheric lightning is perhaps the most widely observed example of this. The electric field becomes so high that the localized electrons in the material become delocalized so that they are free to conduct. This is an important consideration in many situations. For example, the breakdown field of the dielectric insulation around electric cables, which is commonly a plastic material for flexibility, limits the maximum voltage that can be applied. Cables have to be appropriately engineered for a given application, both in terms of materials and geometry.

The metal oxide semiconductor field effect transistor (MOSFET), which is the building block behind microprocessors and memory devices, also relies on dielectrics in its design. The MOSFET is a three-terminal device in which a voltage applied to a gate terminal is used to control the flow of current between the other two terminals, which are known as the source and drain. The device structure shown in Figure 2.3 achieves this. The gate, which is made from a conducting material, is electrically isolated from a semiconducting material that connects the drain and source. Applying a voltage to the gate induces a charge on the gate and an opposite charge in the semiconductor, and it is the modulation of this charge in the semiconductor which regulates the flow of current between the drain and source when

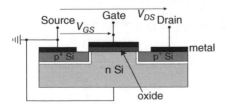

Figure 2.3 A schematic diagram of the cross section of a MOSFET showing the three terminals: the conducting gate, and the source and drain which are connected to a semiconductor. In this case, heavily doped p-type silicon (p^+ Si) and n-type silicon (n Si) are indicated.

a voltage is applied between these two terminals. It is important that the current flow in the device is dominated by that between the drain and source and is not significantly due to any current across the dielectric from the gate. Therefore, it is vital that the gate does not break down. However, to increase processing power in a microprocessor or the capacity of a memory, it is necessary to decrease the physical size of each of the transistors. As the gate dielectric is made thinner, so the voltages applied to the device have to be reduced to avoid breakdown of the dielectric. In practice, dielectrics will conduct at much lower fields than the breakdown field either due to quantum mechanical tunnelling through the dielectric, tunnelling of electrons between localized states in the dielectric or even an ohmic conduction due to a small number of free charge carriers that are present. There is therefore interest in finding materials with high relative permittivity, as increasing the permittivity allows a lower voltage to be used to induce the same charge in the semiconductor and therefore the same switching effect in the transistor, but with lower electric fields (Lo et al. 1997).

While we have considered polarization in a dielectric to be due to the displacement of the negative and positive charge on an atomic scale, other processes can also take place, such as deformation of an ionic lattice or rotation of molecules with an inherent dipole moment. *Liquid crystals* are materials in that consists of long chain molecules which, although in a liquid state, are ordered with the same orientation within a macroscopic region of space called a *domain*. A consequence of this is that the relative permittivity is not isotropic, but is dependent on the direction in which an electric field is acting relative to the direction of orientation. A similar anisotropy in relative permittivity also exists in ionic crystalline materials that are not based upon cubic lattices with high degrees of symmetry. Many naturally occurring crystals display this behaviour, including sapphire, ruby, quartz and mica.

In these cases, we cannot express the relationship between the electric flux density and the electric field as a simple number, but we must use a tensor of the form

$$\begin{pmatrix} D_x \\ D_y \\ D_z \end{pmatrix} = \begin{pmatrix} \varepsilon_{xx} & \varepsilon_{xy} & \varepsilon_{xz} \\ \varepsilon_{yx} & \varepsilon_{yy} & \varepsilon_{yz} \\ \varepsilon_{zx} & \varepsilon_{zy} & \varepsilon_{zz} \end{pmatrix} \begin{pmatrix} E_x \\ E_y \\ E_z \end{pmatrix} \tag{2.20}$$

where the electric flux density in each of three Cartesian axis directions is given by D_x, D_y and D_z, with similar expressions for the electric field in the same three directions. By aligning the Cartesian axes with the orientation of the crystal, the relative permittivity tensor normally diagonalizes to the form

$$\begin{pmatrix} D_x \\ D_y \\ D_z \end{pmatrix} = \begin{pmatrix} \varepsilon_{xx} & 0 & 0 \\ 0 & \varepsilon_{yy} & 0 \\ 0 & 0 & \varepsilon_{zz} \end{pmatrix} \begin{pmatrix} E_x \\ E_y \\ E_z \end{pmatrix} \tag{2.21}$$

Materials where each of ε_{xx}, ε_{yy} and ε_{zz} are different are called *biaxial*, but in many cases the symmetry of the material means that $\varepsilon_{yy} = \varepsilon_{zz}$, and these are called *uniaxial*. Liquid crystals are uniaxial as there is symmetry in the plane perpendicular to the axis of orientation of the liquid crystal molecules. It is for this reason that liquid crystals can be used with polarizers. Polarizers are thin layers of crystalline material that only allow light with an electric field pointing in a specific direction defined by the crystal orientation to pass through. Combining polarizers with liquid crystals allows the transmission of light to be modulated with time under the action of an applied electric field to the liquid crystal.

This is the basis behind the operation of the *liquid crystal display*, which is the technology that has enabled the development of flat panel displays for laptops, smartphones, desktop monitors and flat screen televisions (Chen et al. 2012).

2.3 Capacitors and Energy Storage

Perhaps the most basic electronic device which employs dielectric materials is the *capacitor*. It is a device for storing electrical energy, and it consists of a dielectric material between two conductors. The most basic geometry of this is the *parallel plate capacitor*, shown in Figure 2.4, which is simply two conducting plates of area A with a dielectric of thickness d between them. Application of a voltage V to the two conductors induces equal and opposite charges, $+Q$ and $-Q$, on them; this produces an electric field in the dielectric and so energy is stored in the capacitor. The electric field is confined between the two plates as lines of electric field begin on positive charges and end on negative charges. If the size of the plates is much greater than their separation, then the electric field will be uniform throughout the dielectric (i.e. there will be insignificant *fringing* effects of the field spilling out around the edges of the capacitor).

For any induced charge, we can calculate the electric field inside the dielectric using the Gauss law of electrostatics (as expressed for dielectrics in Eq. (2.14)). If we choose a box-shaped Gaussian surface which surrounds the plate with the positive charge, as shown in Figure 2.4, then as the electric field only exists between the plates and passes perpendicularly through the Gaussian surface, Eq. (2.14) simplifies to

$$|\mathbf{D}|A = Q \tag{2.22}$$

By using Eq. (2.15) to substitute electric flux density for electric field, and denoting $E = |\mathbf{E}|$, we have the result that

$$E = \frac{Q}{\varepsilon_0 \varepsilon_r A} \tag{2.23}$$

We can now use Eq. (1.8) to relate the charge to the applied voltage, taking care to use the coordinate system indicated in Figure 2.4 and noting that V is the voltage applied to the left hand plate with respect to the right, as

$$V = -\int_d^0 \frac{Q}{\varepsilon_0 \varepsilon_r A} dx = \frac{Qd}{\varepsilon_0 \varepsilon_r A} \tag{2.24}$$

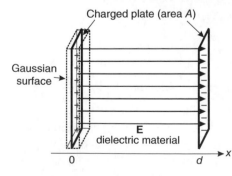

Charged plate (area *A*)

Gaussian surface

E
dielectric material

0   *d*   → *x*

Figure 2.4 A schematic diagram of a parallel plate capacitor consisting of two charged plates of area *A* and separated by a distance *d*, with a positive charge $+Q$ on one plate and a negative charge $-Q$ of the same magnitude on the other, so that an electric field **E** exists in the dielectric between the plates. The Gaussian surface used to calculate **E** is shown.

A linear relationship emerges between the potential difference between the two charged plates and the magnitude of the charge. We therefore define a new quantity, *capacitance*, which is the change in charge induced for a given change in potential difference and is given by

$$C = \frac{dQ}{dV} \tag{2.25}$$

where the linearity of the system means that this simplifies to $dQ/dV = Q/V$. The unit of capacitance is the farad (F). For the case of the parallel plate capacitor, substitution of Eq. (2.25) into Eq. (2.24) produces the result that $C = \varepsilon_0 \varepsilon_r A/d$.

It is also worth noting that applying the chain rule to Eq. (2.25) allows the capacitance to be expressed as

$$C = \frac{dQ}{dt} \frac{dt}{dV}$$

and as current $I = dQ/dt$ this rearranges to

$$I = C\frac{dV}{dt} \tag{2.26}$$

which allows us to directly relate the current charging (or discharging) any capacitor to the rate of change of voltage across the capacitor.

In Section 1.3, we considered that whenever a field occupies a volume of space, energy has been stored. In fact, this energy U can be calculated as

$$U = \int_v \int_0^E \mathbf{D}.d\mathbf{E}dv \tag{2.27}$$

where v is the volume of space in which the field exists. We will return to this when we have put together all of the Maxwell equations in Section 5.6, but for now we can consider this as saying that energy is required to create an electric flux density in a volume of space. The quantity $\mathbf{D}.\,d\mathbf{E}$ is the component of the electric flux density produced in the direction of the applied electric field multiplied by the magnitude of the electric field itself. For a simple dielectric where $\mathbf{D}$ and $\mathbf{E}$ are linearly related by Eq. (2.15), this is the energy stored in the field per unit volume of space, and Eq. (2.27) reduces to

$$U = \int_v \int_0^E \varepsilon_0 \varepsilon_r E dE \, dv = \int_v \frac{1}{2}\varepsilon_0 \varepsilon_r E^2 dv \tag{2.28}$$

This is an extremely powerful expression as it allows us to calculate the energy per unit volume in an electric field occupying a simple dielectric medium.

If we apply Eq. (2.28) to the parallel plate capacitor in Figure 2.4, which has a volume of dielectric $v = Ad$, then by substitution of Eq. (2.23),

$$U = \frac{1}{2}\frac{dQ^2}{\varepsilon_0 \varepsilon_r A} \tag{2.29}$$

and comparison of the result with Eqs. (2.25) and (2.26) gives

$$U = \frac{1}{2}\frac{Q^2}{C} = \frac{1}{2}CV^2 \tag{2.30}$$

We find that these are general expressions for the energy stored in capacitors, whatever their geometry.

The same result can be found by considering the energy required to charge a capacitor up from an initial condition when it has no charge and no potential difference between its two conductors. If we imagine that, at some point during this charging process, the capacitor is at some potential difference V and we add a small charge δQ, then the small change in energy associated with this will be

$$\delta U = V\delta Q \tag{2.31}$$

This will also lead to a small change in potential difference δV, where, from Eq. (2.25),

$$\delta Q = C\delta V \tag{2.32}$$

so Eq. (2.31) can be expressed in terms of potential difference only as

$$\delta U = CV\delta V \tag{2.33}$$

Therefore, the total energy required to charge the capacitor up to some arbitrary potential difference is

$$U = \int_0^V CV dV = \frac{1}{2}CV^2 \tag{2.34}$$

Comparison of Eqs. (2.30) and (2.34) and the methods used to derive them show that the energy required to charge up a capacitor is stored in the form of the electric field that fills the dielectric between the conductors.

2.4 Ferroelectrics

In our consideration of dielectrics in Section 2.2, we assumed that there is a linear relationship between the electric flux density $\mathbf{D}$ and the electric field $\mathbf{E}$ as given by Eq. (2.15) up to the point of dielectric breakdown, as shown in Figure 2.5. In fact, we are assuming that the polarization $\mathbf{P}$ that is induced in the bulk of a dielectric is proportional to the applied electric field, and this is made explicit in Eq. (2.9). For most dielectric materials, such as those where polarization is caused by the displacement of the electrons around nuclei under the influence of the electric field, this is indeed the case. However, in some dielectrics which are made up of ions, the origin of the local dipoles is a distortion of the crystal lattice. An example of such a material is zinc oxide, whose structure is a hexagonal wurtzite lattice of Zn^{2+} and O^{2-} ions. The application of an electric field results in a relative movement of the zinc and oxygen ions within each unit cell, leading to a polarization. Not only can this process lead to a nonlinearity between $\mathbf{D}$ and $\mathbf{E}$, but it is also possible that the distorted lattice is itself stable. Therefore when an applied electric field is removed, the lattice remains in the distorted state, and so a remnant electric flux density will still be present. Cycling the electric field leads to a classic *hysteresis curve* as shown in Figure 2.5.

Such materials are called *ferroelectrics*. As there is no linear relationship between $\mathbf{D}$ and $\mathbf{E}$, it is not possible to define a simple relative permittivity for this class of materials. However, if we were to take a simple ratio of $\mathbf{D}$ and $\mathbf{E}$ and use this to work out an 'effective relative permittivity', then we would calculate values in excess of 1,000 quite widely for many ferroelectric materials, such as lead zirconate titanate (PZT) and barium titanate (Ulrich et al. 2000). This is much higher than the relative permittivity of many non-ferroelectric dielectrics, such as silicon nitride ($\varepsilon_r = 4.2$).

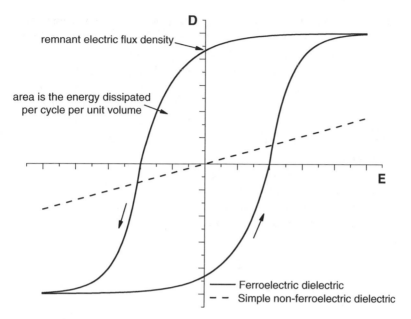

Figure 2.5 The typical hysteresis curve for a ferroelectric material (solid line) compared with the simple linear relationship between electric flux density **D** and electric field **E** for a non-ferroelectric dielectric (dashed line).

According to Eq. (2.27), we can calculate the energy stored in an electric field occupying a volume of space by integrating **D**. **E** as the field is created. Therefore, the area inside the hysteresis curve shown in Figure 2.5 is the energy dissipated per unit volume required to take the ferroelectric material round one cycle, as energy has to be expended to reverse the polarization of the material and take it back to its original polarization again.

One common use of ferroelectrics is in capacitors. The very high effective permittivity of ferroelectrics means that if they are used as the dielectric material between the conductors, then a very large capacitance will result and a very large energy density is stored. Ferroelectrics are also used in memory devices for computing, called ferroelectric random access memories (FeRAMs). As the ferroelectric material can exist stably in one of two polarization states, this can be used to store a binary bit of data ('0' or '1'). Application of a voltage across the ferroelectric material can be used to change the polarization and therefore store data. The polarized ferroelectric can induce a charge in a layer of a semiconductor, and therefore by testing the resistance of this layer of semiconductor, the polarization of the ferroelectric can be deduced and the data read without affecting the data stored (Ishiwara et al. 2004).

2.5 Piezoelectrics

In our discussion of dielectrics in Section 2.2 and ferroelectrics in Section 2.4, we have considered the macroscopic effect of the production of local electric dipole moments in a material under the application of an applied electric field, and we have seen that the formation of the local electric dipole can be as a result of displacement of electrons around nuclei or a distortion of a crystal lattice of ions which result in a relative movement of the positive and negative charge within each unit cell. In the latter case, if a similar distortion of the crystal lattice can be produced mechanically by the application of a stress leading to a strain, then an electric polarization will be produced. This effect where the application of a stress leads to the formation of an electric field is called the *piezoelectric effect*, and the class of materials in which this is observed are called piezoelectrics. Likewise, if an electric field is applied to a piezoelectric material, then the resulting local polarization that is induced causes structural deformation of the lattice and the generation of a mechanical stress, and this is known as the *inverse piezoelectric effect*.

In non-piezoelectric materials, we know that electric flux density and electric field are linked by the relative permittivity, and this is expressed in Eq. (2.20) for an anisotropic material. Furthermore, mechanical stress and strain are linked by the Young modulus of a material, expressed as

$$
\begin{pmatrix} e_x \\ e_y \\ e_z \\ \gamma_{yz} \\ \gamma_{zx} \\ \gamma_{xy} \end{pmatrix} = \begin{pmatrix} S_{11} & S_{12} & S_{13} & S_{14} & S_{15} & S_{16} \\ S_{12} & S_{22} & S_{23} & S_{24} & S_{25} & S_{26} \\ S_{13} & S_{23} & S_{33} & S_{34} & S_{35} & S_{36} \\ S_{14} & S_{24} & S_{34} & S_{44} & S_{45} & S_{46} \\ S_{15} & S_{25} & S_{35} & S_{45} & S_{55} & S_{56} \\ S_{16} & S_{26} & S_{36} & S_{46} & S_{56} & S_{66} \end{pmatrix} \begin{pmatrix} \sigma_x \\ \sigma_y \\ \sigma_z \\ \tau_{yz} \\ \tau_{zx} \\ \tau_{xy} \end{pmatrix} \tag{2.35}
$$

where the e_i are the strain terms, γ_{ij} are the shear strain terms, S_{ij} are the coefficients in the stiffness matrix for the material, σ_i are the stress terms, τ_{ij} are the shear stress terms, and the subscripts x, y and z denote the three crystallographic axes.

In piezoelectric materials, the two independent equations (2.20) and (2.35) become mixed as mechanical deformation affects the electric field and vice versa, giving

$$
\begin{pmatrix} e_x \\ e_y \\ e_z \\ \gamma_{yz} \\ \gamma_{zx} \\ \gamma_{xy} \\ D_x \\ D_y \\ D_z \end{pmatrix} = \begin{pmatrix} S_{11} & S_{12} & S_{13} & S_{14} & S_{15} & S_{16} & d_{11} & d_{12} & d_{13} \\ S_{12} & S_{22} & S_{23} & S_{24} & S_{25} & S_{26} & d_{12} & d_{22} & d_{23} \\ S_{13} & S_{23} & S_{33} & S_{34} & S_{35} & S_{36} & d_{13} & d_{23} & d_{33} \\ S_{14} & S_{24} & S_{34} & S_{44} & S_{45} & S_{46} & d_{14} & d_{24} & d_{34} \\ S_{15} & S_{25} & S_{35} & S_{45} & S_{55} & S_{56} & d_{15} & d_{25} & d_{35} \\ S_{16} & S_{26} & S_{36} & S_{46} & S_{56} & S_{66} & d_{16} & d_{26} & d_{36} \\ d_{11} & d_{12} & d_{13} & d_{14} & d_{15} & d_{16} & \varepsilon_{11} & \varepsilon_{12} & \varepsilon_{13} \\ d_{12} & d_{22} & d_{23} & d_{24} & d_{25} & d_{26} & \varepsilon_{12} & \varepsilon_{22} & \varepsilon_{23} \\ d_{13} & d_{23} & d_{33} & d_{34} & d_{35} & d_{36} & \varepsilon_{13} & \varepsilon_{23} & \varepsilon_{33} \end{pmatrix} \begin{pmatrix} \sigma_x \\ \sigma_y \\ \sigma_z \\ \tau_{yz} \\ \tau_{zx} \\ \tau_{xy} \\ E_x \\ E_y \\ E_z \end{pmatrix} \tag{2.36}
$$

The piezoelectric constants d_{ij} have units of C N^{-1} and provide the link between the electrical and mechanical response of the piezoelectric material.

Some materials, such as zinc oxide or aluminium nitride, have a well-defined crystallographic orientation and so always show a piezoelectric response. In other materials, such as polyvinylidene fluoride (PVDF), there is naturally a random orientation of the electric dipoles inside the material and so no macroscopic piezoelectric effect is observed until these dipoles are aligned by application of a very large electric field to the material, in a process known as *poling*.

2.6   Metals

Thus far, we have only considered the interaction of electric fields with materials that contain no free charges. In metals, this is not the case. Bonding in metallic materials is characterized by the delocalization of the valence electrons associated with atoms, and these delocalized electrons are free to move over macroscopic distances. In aluminium, for example, which is a Group III element, each atom has three valence electrons, all of which delocalize at room temperature with the result that there is a number density of $\sim 1.8 \times 10^{29}$ m^{-3}.

Therefore, if we imagine applying an external electric field to a block of metallic material, as shown in Figure 2.6, then the electrons in the metal will be attracted towards the surface of the metal where they will accumulate a net negative charge and will be repelled from the opposite surface leaving a positive surface charge, until the electric field is entirely blocked from entering the metal. The vast number density of electrons in the metal ensures that no static electric field can penetrate.

The inability of static electric fields to penetrate metals has an important consequence that, if a metal does have a net charge for any reason, then this charge can only exist on the surface of the conductor. It is a common misconception that net charge only exists on the surface of metals because the charges all repel each other, and this gets them as far away from each other as possible. This does not make sense, however – the atoms in an ideal gas want to be as far apart from each other as possible according to the kinetic theory of gases, but the gas molecules inside an inflated balloon do not all stick to the inner surface! In order to appreciate the correct physics behind this effect, let us consider a coaxial cable (such as is used to bring the electrical signal from a television aerial on the roof of a house to a television set).

The coaxial cable consists of a solid, central metallic conductor which is surrounded by a layer of flexible dielectric material, such as polytetrafluoroethylene (PTFE), which is itself surrounded by a layer of braided metal (which we will consider behaves like a solid metal). The whole cable is coated in a plastic sheath, as shown in Figure 2.7. We will imagine that we have a length l of the cable, and we will apply a d.c. voltage source V to one end of the cable. The cable behaves as a capacitor as we have two conductors separated by a layer of insulating dielectric, and so a charge transfer will take place between the two conductors through the voltage source which will result in a net positive charge $+q$ on the inner

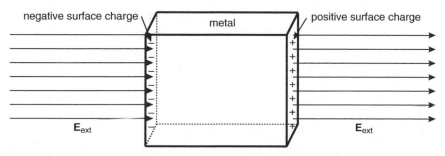

Figure 2.6   The effect of an applied external electric field $\mathbf{E}_{ext}$ on a block of metal is to induce a charge on its surface such that the inside of the metal is completely screened from the applied field.

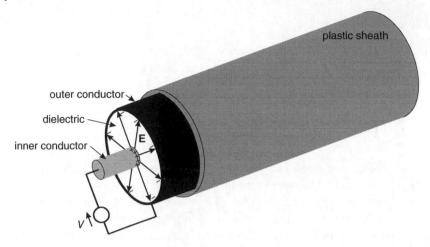

Figure 2.7 Cross-section of a coaxial cable with a potential difference *V* applied between the inner and outer conductors. This results in a positive charge being developed on the outer surface of the inner conductor and a negative charge on the inner surface of the outer conductor. An electric field **E** therefore exists in the dielectric between the conductors, but this does not penetrate into either metal.

conductor and a net negative charge $-q$ on the outer conductor. Where exactly will these two net charges exist?

One way of answering this question is to think of this as a 'Romeo and Juliet' problem. In the famous scene in Shakespeare's play, Juliet is on a balcony and Romeo is underneath. They are separated by the air gap between them, but because they are attracted to each other, they get as close to each other on either side of the gap as possible. So it is for the positive and negative charge. They are strongly attracted to each other but are separated by the dielectric, and so they get as spatially close to each other as possible. Therefore, the positive charge must exist on the outer surface of the inner conductor, and the negative charge on the inner surface of the outer conductor. Remembering that lines of electric field begin on positive charges and end on negative charges and that $|+q| = |-q|$, it is clear from Figure 2.7 that no electric field penetrates either metal, as would be expected. We can check for consistency of this argument using the Gauss law of electrostatics by considering a small Gaussian surface which is entirely contained within the central conductor. If there was an electric field inside the metal, then the free electrons would experience a force and would move to screen the electric field. Therefore, inside the metal $\mathbf{E} = \mathbf{0}$, and hence $\oint_S \mathbf{E} \cdot d\mathbf{A} = 0$, which means that there can be no net charge inside the Gaussian surface according to Eq. (1.12). This proves that as there is no electric field inside the metal, there can be no net charge inside the metal either – charge can only accumulate on the surface.

Another consequence of there being no electric field inside a metal follows from Eq. (1.9), which states that the potential difference between two points in space (*a* and *b*) is $V_{ab} = -\int_b^a \mathbf{E} \cdot d\mathbf{x}$. Hence, if $\mathbf{E} = \mathbf{0}$, then the integral can only give a zero result, and therefore all points inside a block of metal are at the same equipotential. Intuitively, we know that this is true, because if there was a potential difference inside the metal, then the free electrons would flow down the potential difference.

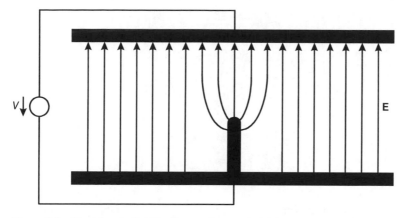

Figure 2.8 The electric field is distorted around a high aspect ratio structure placed in a region of uniform electric field, such as a pair of parallel conducting plates with a potential difference *V* applied between them. The requirement that the electric field lines intersect perpendicularly with the metal surface leads to an increased local electric field.

We saw in Section 1.3 that lines of electric flux always cut perpendicularly through lines of equipotential. Therefore, if the surface of a metal is at a constant equipotential, then lines of electric field must meet the surface of a metal perpendicularly. This is clearly the case for the coaxial cable in Figure 2.7. We can use this effect to engineer a locally very high electric field by creating a high aspect ratio structure on an otherwise flat, metallic surface, as shown in Figure 2.8. In this case the requirement that the electric field is perpendicular to the surface means that the field lines are distorted around the high aspect ratio structure to produce a very high electric field.

On a macroscopic scale, this is why is it dangerous to play golf in a thunderstorm! Golf courses are typically flat, open areas, and a golf club is a high aspect ratio structure. Therefore, holding the club in the air increases the electric field locally and makes it more likely that a lightning strike will occur.

On a microscopic scale, carbon nanotubes, such as those shown in Figure 2.9, are also high aspect ratio structures and can be highly conducting, depending on their atomic structure (Teo et al. 2003). By applying a voltage to a carbon nanotube with respect to a counter electrode some distance away, a very high electric field can be produced at the tip which can lead to the field emission (spontaneous emission) of electrons from the tip to produce an electron beam. Beams of electrons like this are needed for applications including electron microscopes, some video displays and X-ray sources.

We have seen that we cannot get an electric field that is applied externally to a metal to penetrate inside, but we can get an electric field to exist inside a metal when we cause a current to flow through it. Let us consider the situation shown in Figure 2.10 where we have a battery connected to either end of a metallic wire. The battery has a potential difference *V* between its terminals as the chemical reactions taking place on either side of the battery cause electrons on the side of the negative terminal to have a higher potential energy than electrons on the side of the positive terminal. As there are many free electrons inside the metal wire which can move down the potential difference from the negative to the positive terminal, we know that a 'classical' current *I* will appear to flow from the positive to the

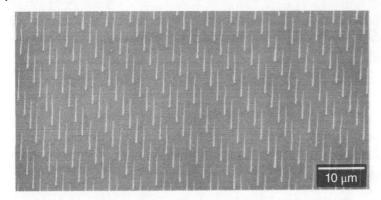

Figure 2.9 Scanning electron microscopy image of an array of carbon nanotubes on the surface of a silicon wafer. Source: Reprinted with permission from K. Teo *et al.* (2003), Fabrication and electrical characteristics of carbon nanotube-based microcathodes for use in a parallel electron-beam lithography system. *Journal of Vacuum Science and Technology B* **21**(2): 693–697. Copyright 2003, American Vacuum Society.

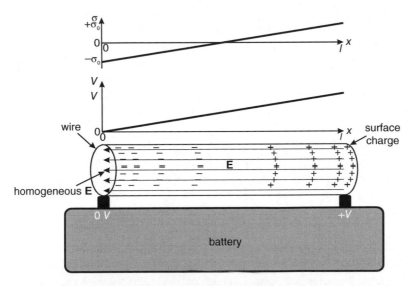

Figure 2.10 Schematic diagram of a metallic wire of length l connected to a battery with a potential difference V between its terminals. A surface charge is developed on the wire which produces a uniform electric field E inside the wire and therefore a linear variation in potential along the wire. Graphs of both surface charge density and potential are shown as a function of position x along the wire.

negative terminal. This current will be finite, as the electrons in the wire will be scattered by impurities and defects in the wire, and by the thermal vibration of the positively charged ions in the metal lattice. The *Ohm law* quantifies this by defining a *resistance* of the wire R with units of ohms (Ω) such that

$$R = \frac{V}{I} \tag{2.37}$$

As R will depend on the geometry of the wire, it is helpful to also define a geometry-independent material property called the *resistivity* ρ which is given by

$$\rho = \frac{RA}{l} \qquad (2.38)$$

where A and l are the cross-sectional area and length of the wire, respectively. The units of resistivity are Ω m.

The flow of a simple d.c. current in a metal can be considered to be like the flow of an incompressible fluid, since if we could compress the free charge, then this would result in a net charge in a small volume of space which would set up a locally inhomogeneous electric field to make the charge spread out again to restore charge neutrality. This ensures that the same current I will be flowing at all points in the wire. If a junction is introduced into the circuit, as is the case in Figure 2.11, then the total current flowing into the junction must equal the total current flowing out. This is the *Kirchhoff current law*, and it can be expressed as

$$\sum I_{out} = 0 \qquad (2.39)$$

where $\sum I_{out}$ is the sum of the currents flowing out along all possible conduction paths from a point (considering a current flowing in to be a negative current flowing out). In essence, this is an expression of the *conservation of charge*.

Although there are no locally inhomogeneous electric fields inside a metal wire carrying a current, there must be a homogeneous electric field which points axially along the length of the wire so that, at any point in the bulk of the wire, an electron will experience a force, causing it to move. As the same current is flowing at all points in the wire if it is of uniform cross section and resistivity, the axial electric field must have the same magnitude at all points in the wire and hence, as $E = -dV/dx$ from Eq. (1.6), there must be a constant gradient in the potential difference along the wire. In other words, the potential at any distance x along the wire away from the negative terminal must be

$$V(x) = \frac{Vx}{l} \qquad (2.40)$$

where V is the potential difference across the battery and l is the length of the wire. This has to be the case, as we saw in Eq. (1.11) that the integral of electric field around a closed loop $\oint_C \mathbf{E}. \, \mathbf{dx} = 0$ to ensure *conservation of energy*. For the case of a closed electrical

Figure 2.11 The Kirchhoff Current Law states that the sum of all of the currents out of a point in a circuit must be zero due to conservation of charge.

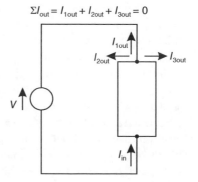

circuit containing a number of elements, with each element having a potential difference V across it,

$$\sum V = 0 \qquad (2.41)$$

where $\sum V$ is the sum of the potential differences around a whole loop of the circuit (Kirchhoff 1850). This is the *Kirchhoff voltage law*.

There remains an outstanding question, however. If the bulk of a metal is electrically neutral, where is the charge that produces the homogeneous axial electric field inside a metal wire causing the current to flow? Although Gustav Kirchhoff is best known for his work in the mid-nineteenth century on the two circuit laws that bear his name, he also considered this fundamental issue (Kirchhoff 1850; Kirchhoff 1857). The application of the potential difference between the two ends of the wire by the battery causes a layer of charge to form on the surface of the wire, as shown in Figure 2.10. As the wire had no net charge before the battery was connected to it, this must still be the case after connection of the battery, and so the integral of this surface charge over the whole length of the wire will be zero. The surface charge density will be $-\sigma_0$ on the wire at the end connected to the negative terminal of the battery and will be $+\sigma_0$ at the end connected to the positive terminal, varying linearly between the two, having the form

$$\sigma(x) = \frac{2\sigma_0 x}{l} - \sigma_0 \qquad (2.42)$$

(assuming the wire to be uniform along its length). This linearly varying charge density will produce the homogeneous axial electric field inside the wire and the linearly varying potential of Eq. (2.40). It should be noted that this surface charge will produce an electric field around the outside of the wire as well as the internal field. For a quantitative discussion of the electric field and surface charge for a particular wire geometry as an example of specific forms of these equations, the reader is referred to the work by Assis et al. (1999).

2.7 Semiconductors

The final class of materials for which we will consider the interaction with electric fields is semiconductors. Semiconductors have enabled the microelectronics revolution that has transformed the way that we live since the 1950s. The operation of transistors is entirely dependent on semiconductors, and these devices are the essential elements of both microprocessors and memories. Semiconductors are also needed to make light-emitting diodes, solid-state lasers and photodiodes, which are all elements of the optical fibre-based technology upon which high-speed communications depend.

Semiconductors are distinct from either metals or dielectrics, and to understand why this is, we shall consider the case of silicon, which is the dominant semiconductor material. Silicon is a Group IV element which takes up a covalent bonding structure in which valence electrons are shared between neighbouring atoms. Most covalently bonded materials are highly resistive dielectrics, as the electrons in the covalent bonds are held in deep potential wells, and so there are no free electrons to carry a current. However, in semiconductors such as silicon, the potential well associated with the covalent bonds is sufficiently low (1.1 eV for silicon) that, at room temperature, there is a small but finite probability (around 10^{-12} for silicon) that an electron will have enough thermal energy to escape from the covalent bond and become free to move either by *diffusion* or under the influence of an electric field (called *drift*). This results in a number density $\sim 10^{16}$ m^{-3} of free electrons in silicon (compared with number densities of free electrons $\sim 10^{29}$ m^{-3} in metals). The covalent bond from which a free electron has escaped is now short of an electron; this absence of an electron can also move by drift or diffusion, and appears to behave like a free positive charge which is called a *hole*.

The precise number of free electrons and holes in a semiconductor can be engineered through the addition of impurities, called *dopants*. For example, the number of free electrons can be increased by introducing a small quantity of a Group V element into silicon, such as phosphorus. The bonding configuration of the phosphorus impurity is such that it is very easily ionized, producing a free electron and leaving behind a positively charged phosphorus ion, but without the formation of a hole. Free holes can be produced in silicon by adding a Group III impurity, such as boron, which again is ionized. This time a free hole is produced, but with no free electron, and the boron becomes a negatively charged ion. To a very good first approximation, the number density of free electrons or holes at room temperature is equal to the number density of impurities that were added, which is normally no more than 10^{23} m^{-3} for silicon.

Although silicon is the most commonly used semiconductor, there are a wide diversity of semiconducting materials. Many, such as Ge, GaAs, and InP, are covalently bonded, but some, such as ZnO, are ionic. Whatever the precise bonding mechanism, in all cases, there is a small energy barrier to the thermal creation of free electrons and holes, and the number density of free electrons and holes can be controlled through the addition of ionized impurity dopants. Semiconductors where the majority of free carriers are electrons are called *n-type* and those where the majority of free carriers are holes are called *p-type*.

The consequence of semiconductors having small and well-defined number densities of electrons or holes in their bulk created by ionized dopants is that, unlike in metals, static electric fields are able to penetrate into semiconductors and so change the local number density of free electrons or holes, but also, unlike in dielectrics, the depth of penetration of the field is limited.

For example, let us imagine that we create a capacitor structure consisting of an insulating layer of dielectric material of thickness d_i with a metal on one side and a p-type semiconductor on the other, as shown in Figure 2.12. If silicon dioxide is used as the insulator, then this is commonly called a *metal-oxide-semiconductor* (MOS) structure, and it is effectively the structure of a MOS field effect transistor (MOSFET) at the gate of the device, as shown in Figure 2.3. We will imagine that we have applied a voltage V to the metal with respect to the bulk of the semiconductor. As a result, a charge per unit area, $+\sigma$, is formed on the lower surface of the metal through the local movement of electrons away from the surface. A balancing charge per unit area of $-\sigma$ must form in the semiconductor through the repulsion of holes away from the surface. This leaves behind the negatively charged dopant ions, which are present with a finite number density N per unit volume. This is called a *depletion region*, as the volume has been depleted of free holes which could conduct a current. In

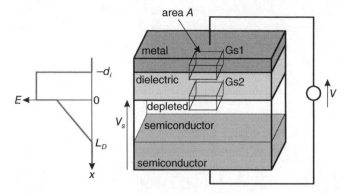

Figure 2.12 Schematic diagram of the metal oxide semiconductor capacitor structure. Gs1 and Gs2 are Gaussian surfaces 1 and 2 respectively. Both the depleted and 'undepleted' regions of semiconductor materials are identified. The electric field as a function of position through the structure is shown on the left-hand side of the diagram.

order for there to be a charge per unit area of $-\sigma$, this depletion region must extend a finite depth L_D into the semiconductor, as indicated in Figure 2.12, so that

$$-eNL_D = -\sigma \tag{2.43}$$

as $-eN$ must be the charge per unit volume in the depletion region due to the ionized dopants. As there is a charge present, there must also be an electric flux density of magnitude D_i in the insulating dielectric and magnitude D_s in the semiconductor.

The former of these can be easily evaluated using the Gauss Law of electrostatics (Eq. (2.14)), using Gaussian surface 1 (Gs1) in Figure 2.12, as

$$\oint_S \mathbf{D.dA} = \sigma A$$

$$D_i = eNL_D \tag{2.44}$$

where $D_i = |\mathbf{D}_i|$ and a substitution has been made using Eq. (2.43). If the relative permittivity of the insulating dielectric is ε_{ri}, then the magnitude of the electric field in the dielectric is

$$E_i = \frac{eNL_D}{\varepsilon_0 \varepsilon_{ri}} \tag{2.45}$$

and is clearly uniform throughout the entire volume of the dielectric.

Gaussian surface 2 (Gs2) allows the electric flux density in the semiconductor to be evaluated. This is slightly complicated by the presence of the charge density due to ionized dopants which means that the flux density will vary with depth x into the semiconductor. However, by considering Gaussian surface 2 it can be seen that

$$\oint_S \mathbf{D.dA} = -eNxA$$

$$D_s - D_i = -eNx \tag{2.46}$$

By substituting for D_i in Eq. (2.44) and rearranging, we have

$$D_s = eN(L_D - x) \tag{2.47}$$

and if the relative permittivity of the semiconductor is ε_{rs}, then the magnitude of the electric field as a function of depth into the semiconductor can be expressed as

$$E_s = \frac{eN(L_D - x)}{\varepsilon_0 \varepsilon_{rs}} \tag{2.48}$$

which is valid for $0 \leq x \leq L_D$. In other words, the electric field is a maximum at the surface of the semiconductor, and decays linearly to zero through the depletion region, as indicated in Figure 2.12.

We can now use Eqs. (1.8) and (2.48) to evaluate the potential on the surface of the semi-conductor with respect to the bulk V_s,

$$V_s = -\int_{L_D}^{0} \frac{eN(L_D - x)}{\varepsilon_0 \varepsilon_{rs}} dx = \frac{eNL_D^2}{2\varepsilon_0 \varepsilon_{rs}} \tag{2.49}$$

which rearranges to

$$L_D = \sqrt{\frac{2\varepsilon_0 \varepsilon_{rs} V_s}{eN}} \tag{2.50}$$

Therefore, by applying a potential difference to the metal with respect to the semiconductor, we can control the surface potential V_s of the semiconductor and hence the thickness of the depletion region that is formed. By way of an example, for silicon with a density of dopants of 10^{22} m^{-3} and a surface potential of 1 V, the depletion region will extend ~350 nm into the semiconductor.

As the depletion region is insulating in nature, when such a structure is used in the MOS-FET of Figure 2.3, the voltage on the gate controls the resistance of the semiconductor between the source and drain, and hence the amount of current that will flow between them for a particular potential difference between the source and drain. This is the basis of the operation of the device. By putting MOSFETs into simple circuits, it is possible to make *logic gates* (AND, OR, NOT, etc.) which is the basis behind microprocessors. Alternatively, memories can be made by storing charge on the gate and then using the amount of current flowing between the source and drain to check whether or not charge has been stored, and hence whether a logical '0' or '1' is being stored.

It is this ability for electric fields to penetrate into semiconductors and hence change the number of electrons and holes that are present in a finite volume of material which is key to the operation of semiconductor devices.

References

Assis, A.K.T., Rodrigues, W.A. Jr., et al. (1999). The electric field outside a stationary resistive wire carrying a constant current. *Foundations of Physics* 29 (5): 729–753.

Chen, J., Cranton, W.M. et al. (ed.) (2012). *Handbook of Visual Display Technology*. Berlin: Springer.

Lo, S.H., Buchanan, D.A. et al. (1997). Quantum-mechanical modeling of electron tunneling current from the inversion layer of ultra-thin-oxide nMOSFET's. *IEEE Electron Device Letters* 18 (5): 209–211.

Ishiwara, H., Okuyama, M. et al. (2004). *Ferroelectric Random Access Memories : Fundamentals and Applications*. Berlin: Springer.

Kirchhoff, G. (1850). On a deduction of Ohm's laws, in connexion with the theory of electro-statics. *Philosophical Magazine Series 3* 37 (252): 463–468.

Kirchhoff, G. (1857). On the motion of electricity in wires. *Philosophical Magazine Series 4* 13 (88): 393–412.

Teo, K., Chhowalla, M. et al. (2003). Fabrication and electrical characteristics of carbon nanotube-based microcathodes for use in a parallel electron-beam lithography system. *Journal of Vacuum Science and Technology B* 21 (2): 693–697.

Ulrich, R., Schaper, L. et al. (2000). Comparison of paraelectric and ferroelectric materials for applications as dielectrics in thin film integrated capacitors. *The International Journal of Microcircuits and Electronic Packaging* 23 (2): 172–180.

3

Currents and Magnetic Fields

3.1 Moving Charges and Forces

When we considered the origin of the electrostatic force in Section 1.1, we used the fact that charge is a fundamental property of matter, rather like gravity, and that charge produces an electric field in which another charge experiences a force. It would be helpful if this chapter on magnetic fields could begin with such a simple equivalent statement for magnetic fields, but the origin of magnetic fields is not quite so straightforward.

What is known experimentally is that a charge q_2 which is moving in an inertial frame of reference will experience not only an electrostatic force, as discussed in Section 1.2, but also an additional force in the presence of another charge q_1 which is *also* moving in the same reference frame. It is this additional force which we think of as being magnetism.

Magnetism is actually the result of a combination of electrostatics and *special relativity*. In fact, it was partly as a result of trying to understand magnetism that Albert Einstein developed his special theory of relativity (Einstein 1905). It is not the purpose of this book to provide the reader with a derivation of magnetism from electrostatics using special relativity, as an engineer does not require this knowledge. However, it is useful to have a rough appreciation of the origin of magnetic effects.

At the heart of special relativity is the postulate that if we move from one inertial frame of reference to another that is moving at a different velocity, then space will appear to be distorted in dimensions, such that the distance between two objects appears to change. This distortion also causes forces to change between frames of reference as well. Now consider again our two moving charges, q_1 and q_2. If we are in the frame of reference of charge q_1, then q_2 will appear to be moving with some velocity $\mathbf{v}_2$, whereas if we are in the frame of reference of q_2 then q_1 will appear to be moving with a velocity $\mathbf{v}_1$ such that $\mathbf{v}_2 = -\mathbf{v}_1$. These two frames of reference are effectively special cases, as the relativistic distortion of one charge with respect to the other in both cases is the same, and the same simple electrostatic force acts on both charges. However, in any other inertial frame of reference, a different relativistic distortion will apply to each of the two charges due to their different velocities, with the result that a different force acts on the two charges compared with the previous special cases. It is this difference in the force that actually acts between the two moving charges compared with the simple electrostatic force that we consider to be the magnetic effect.

Electromagnetism for Engineers, First Edition. Andrew J. Flewitt.
© 2023 John Wiley & Sons Ltd. Published 2023 by John Wiley & Sons Ltd.
Companion website: www.wiley.com/go/flewitt/electromagnetism

Happily, if we are content to say that a moving charge produces a magnetic field in which a second moving charge experiences a force, then we do not need to explicitly consider relativistic effects further.

3.2 Magnetic Fields and Moving Charges

Let us consider further the magnetic field that is produced by moving charges using the simplest possible case of a charge q_1 moving at a velocity $\mathbf{v}_1$ in free space, as shown in Figure 3.1. We know that another charge q_2 moving with a non-zero velocity $\mathbf{v}_2$, whose instantaneous position in space is defined by a vector $\mathbf{r}$ with respect to charge q_1, will experience a magnetic force. In Section 1.2, we saw that fields are used to describe regions of space in which objects with a particular property experience a force. Therefore, we say that a *magnetic flux density* $\mathbf{B}$ exists around the moving charge, which is quantitatively given by

$$\mathbf{B} = \frac{\mu_0}{4\pi} \frac{q_1 \mathbf{v}_1 \times \mathbf{r}}{|\mathbf{r}|^3} \tag{3.1}$$

where μ_0 is a constant called the *permeability of free space*. We will not attempt to derive this here from special relativity. Figure 3.1 attempts to show the magnetic flux lines in space around the moving charge q_1. The lines of magnetic flux form closed loops which circulate around the direction of motion of the charge. The direction of circulation is clockwise if q_1 is positive and being viewed from a point behind the direction of motion. This can be remembered using the *right-hand rule*: if the thumb of your right hand points in the direction that the positive charge is moving in then the fingers of your right hand will curl round in the direction of circulation of the magnetic flux.

The magnetic flux density is strongest in a plane which contains the charge q_1 and is perpendicular to $\mathbf{v}_1$, and drops to zero along the line of $\mathbf{v}_1$. Also, noting that the position vector $\mathbf{r}$ occurs in the numerator of Eq. (3.1), the magnetic flux density tends to decrease with distance from the moving charge as $|\mathbf{r}|^{-2}$. In fact, if we define the angle between $\mathbf{r}$ and $\mathbf{v}_1$ as θ, then using the definition of the cross product, we can get a simple expression for the magnitude of the magnetic flux density around the moving charge as

$$B = \frac{\mu_0 q_1 |\mathbf{v}_1| \sin\theta}{4\pi |\mathbf{r}|^2} \tag{3.2}$$

The force which then acts on charge q_2 due to the magnetic field produced by q_1 is given by the *Lorentz force law*,

$$\mathbf{F} = q_2 \mathbf{v}_2 \times \mathbf{B} \tag{3.3}$$

Therefore, if both q_1 and q_2 are both positive and are moving parallel to each other in the same direction, then the (magnetic) Lorentz force will act to attract q_2 toward the line of motion of q_1, as shown in Figure 3.1.

In practice, the charge q_2 will experience both an electrostatic force and the Lorentz force, and so we can combine Eq. (3.3). with Eq. (1.1) to give the total force acting on a charge in the presence of both a magnetic flux density and electric field as

$$\mathbf{F} = q(\mathbf{E} + \mathbf{v} \times \mathbf{B}) \tag{3.4}$$

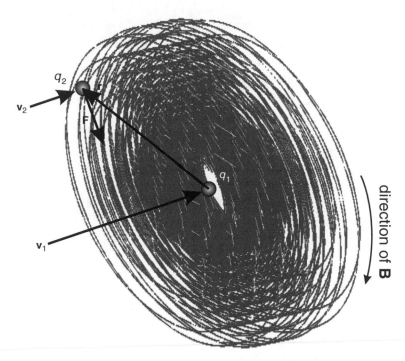

direction of **B**

Figure 3.1 The magnetic flux lines surrounding a positive point charge q_1 moving with a velocity $\mathbf{v}_1$ in free-space. A second positive point charge q_2 moving with a velocity $\mathbf{v}_2$ in this field experiences a force **F**.

For the specific situation shown in Figure 3.1, we can substitute into Eq. (3.4) expressions for both the electric field and magnetic flux density to give

$$\mathbf{F} - \frac{q_1 q_2}{4\pi |\mathbf{r}|^3} \left(\frac{\mathbf{r}}{\varepsilon_0} + \mu_0 \mathbf{v}_2 \times (\mathbf{v}_1 \times \mathbf{r}) \right) \tag{3.5}$$

Again, if we take a special case where $\mathbf{v}_2 = \mathbf{v}_1$, then using the triple cross product rule that $\mathbf{A} \times (\mathbf{B} \times \mathbf{C}) = (\mathbf{A} . \mathbf{C})\mathbf{B} - (\mathbf{A} . \mathbf{B})\mathbf{C}$, it can be shown that $\mathbf{v}_2 \times (\mathbf{v}_1 \times \mathbf{r}) = -|v_1|^2 \mathbf{r}$, so both the electrostatic and Lorentz forces act along the line of **r**, but in opposite directions. Also, the magnitude of the Lorentz force expressed as a fraction of the electrostatic force is $|v_1|^2 \mu_0 \varepsilon_0$. We shall see in Section 7.3 that $1/\sqrt{\mu_0 \varepsilon_0}$ is equal to the speed of light in free space, c, and therefore the magnitude of the Lorentz force in this special case is $|v_1|^2/c^2$, and so is always less than the electrostatic force, but the two would tend towards being equal if the velocity of the charges did approach the speed of light, which is another indication of the relativistic origin of magnetic effects (see Section 3.1).

3.3   The Ampère Circuital Law in Free Space

In Section 3.2, we have seen that magnetic fields are produced by moving charges. Specifically, lines of magnetic flux form closed loops that circulate around moving charges.

This is the basis behind the Ampère circuital law for magnetic fields, which states that

$$\oint_C \mathbf{B} \cdot \mathbf{dr} = \mu_0 I \qquad (3.6)$$

where the left-hand side is an integral around a closed *Amperian loop*, and I is the total current flowing through the loop from the front to the back when viewed such that the integration around the Amperian loop is clockwise. We will now consider the mathematical basis for this equation.

Let us imagine that we have an arbitrary line in three dimensions, such as that in Figure 3.2. We could work out the length of the line by straightening it out and simply measuring its length, but a more elegant approach would be to divide the line up into small elements, so that to a good approximation each of these elements is a straight line which can be defined by a simple vector line element $\mathbf{dr}$. We could then calculate the length of the line by simply integrating the magnitude of these $\mathbf{dr}$ vectors along the whole length of the line:

$$\int_{\text{line}} |\mathbf{dr}| = \text{total length} \qquad (3.7)$$

If there is a magnetic flux density $\mathbf{B}$ present in the same region of space as a line element $\mathbf{dr}$, then $\mathbf{B} \cdot \mathbf{dr}$ is the component of the magnetic field acting in the direction of the line multiplied by the line length. We now need to define an imaginary closed loop, called an Amperian loop which we shall use simply as a mathematical construction for the purpose of calculating the magnetic flux density, rather as we used the Gaussian surface for evaluating electric fields (see Section 1.4). Therefore, if we have a closed Amperian loop around which we integrate $\mathbf{B} \cdot \mathbf{dr}$, then the result will be a measure of the extent to which the magnetic field is circulating in the same direction as the Amperian loop. To exemplify the significance of this, let us imagine that we have a long, straight wire carrying a current I in free space. We know that the magnetic flux lines will circulate in closed loops around the wire as shown in Figure 3.3. If we integrate $\mathbf{B} \cdot \mathbf{dr}$ around loop A in the figure, then far from the wire we find that $\mathbf{B}$ and $\mathbf{dr}$ have an angle between them that is less than 90° and the dot product will be positive, but close to the wire the angle is greater than 90°, and so the dot product will be negative. The total integral around the loop will then be zero. The only way that $\oint_C \mathbf{B} \cdot \mathbf{dr}$ can be non-zero is if $\mathbf{B}$ and $\mathbf{dr}$ point in broadly the same direction all the way around the loop,

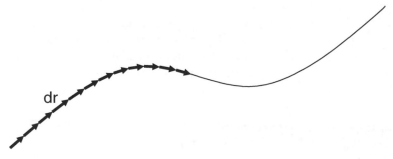

dr

Figure 3.2 An arbitrary line can be divided up into small vector line elements **dr**. Integration of the magnitude of all of the vector elements along the whole line would give the total length, as expressed in Eq. (3.7).

Figure 3.3 The magnetic flux density **B** around a long, straight wire carrying a current *I*. Two Amperian loops are shown. Integrating around loop A of **B**. **dr** will give zero as no current passes through the loop, whereas integration around loop B will give a non-zero result.

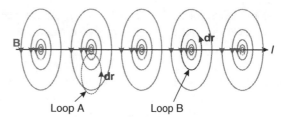

Loop A   Loop B

and as the magnetic flux circulates around currents, this can only be the case if there is a non-zero net current flowing through the loop. This situation is clearly satisfied for loop B in Figure 3.3. The Ampère circuital law (Eq. (3.6)) expresses this mathematically by saying that $\oint_C \mathbf{B}.\, \mathbf{dr}$ is proportional to the current flow through the Amperian loop. The permeability of free space μ_0 is then the constant of proportionality which ensures the correct numerical values.

3.4 Application of the Ampère Circuital Law

Just as the Gauss law of electrostatics allowed us to calculate the electric field produced by a charge distribution (see Section 1.5), so the Ampère circuital law permits the calculation of magnetic fields produced by a current distribution. The line integral of Eq. (3.6) can again seem very off-putting, but good choice of the geometry of the line integral usually makes calculations relatively trivial. The rule of thumb is that an Amperian loop should either follow a flux line so that **B** and **dr** are parallel, thereby reducing the dot product to a simple product of magnetic field and length, or the loop should run perpendicularly to the flux lines to reduce the dot product to zero.

By way of an example, let us consider the magnetic flux density surrounding the simple, long, straight wire shown in Figure 3.3. Loop B is running in a plane perpendicular to the wire, is circular with a radius *R*, and is centred on the wire. Considering the left-hand side of Eq. (3.6), as **B** and **dr** are parallel at all points around the loop, the dot product becomes an ordinary product. Also, due to the radial symmetry of the system, we can see that |**B**| will be constant around the loop. As a current *I* is flowing through the Amperian loop, Eq. (3.6) becomes

$$B \cdot 2\pi R = \mu_0 I \tag{3.8}$$

where $B = |\mathbf{B}|$. This then rearranges simply to

$$B = \frac{\mu_0 I}{2\pi R} \tag{3.9}$$

Therefore, as we would probably expect intuitively, the magnetic field around the wire is proportional to the current and decreases as the inverse of the radial distance from the wire.

If a second, parallel wire running at a distance *R* from the first wire is also carrying a current I_2 in the same direction as *I*, then according to the Lorentz force law (Eq. (3.3)), this second wire will experience a force per unit length *F*. We know that the current in a wire is related to the number density of free electrons N, the area of the wire A, and the average drift velocity of the free electrons v according to

$$I = NAve \tag{3.10}$$

where e is the magnitude of the electronic charge. In order to evaluate the force per unit length acting on the wire, we need to know the total charge of the free electrons in a unit length of the wire, as this will be q_2 in Eq. (3.3). This is the number density of free electrons multiplied by the volume of a unit length of the wire and the electronic charge, which is

$$q_2 = NAe \tag{3.11}$$

Therefore, by substituting Eq. (3.11) into Eq. (3.10), we have that

$$q_2 v = I \tag{3.12}$$

As the magnetic field and the current are perpendicular to each other along the length of the second wire, the cross product in Eq. (3.3) becomes a simple product, so

$$F = BI \tag{3.13}$$

into which we can substitute the equation for the magnetic field (Eq. (3.9)) to give

$$F = \frac{\mu_0 I I_2}{2\pi R} \tag{3.14}$$

The direction of the force is such that the two wires will be attracted together. If the current in the two wires were running antiparallel to each other, then the force would be repulsive. This is consistent with the case for two individual charges that we considered in Section 3.2.

Another common device for producing magnetic fields is a simple coil of wire, which is usually called a *solenoid*, and is shown in Figure 3.4. A coil of wire is a convenient way of producing a magnetic field as it is easy to manufacture and produces a well-defined, and potentially strong, magnetic field through its centre. It is used to make *inductors*, which we will look at in detail in Section 3.8, and *transformers* (see Section 12.3). We will use the Ampère law to calculate the magnetic flux density inside the solenoid. We will assume that the solenoid consists of a spiral of n turns of wire which forms a tube of radius R and length l.

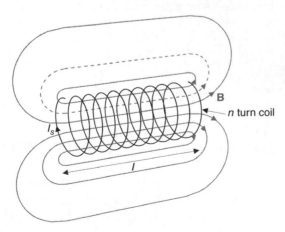

Figure 3.4 A schematic diagram of the magnetic flux density **B** in a plane through a solenoid of n turns and length l carrying a current I_s. One of the flux lines shown dashed can be used as the Ampèrian loop for evaluating the magnetic field inside the solenoid.

A current I_s flows through the coil. Using the right-hand rule (see Section 3.2), it can be seen that the current at any point in the coil acts to produce a magnetic field which runs from left to right through the centre of the solenoid as shown in Figure 3.4. The flux lines then spread out once they leave the right-hand end of the solenoid, but as they must form closed loops, they follow a path back to the left-hand end where they enter the solenoid once more. We will assume that the magnitude of the magnetic field inside the solenoid, B_s, is uniform over the whole cross section.

To evaluate B_s, we will use the Ampère circuital law which requires an Amperian loop. We will use the rule of thumb that it is sensible for the loop to follow a flux line, as shown in Figure 3.4, and then evaluate Eq. (3.6) around this loop. Inside the solenoid, the integral of **B**. **dr** becomes a simple product of magnetic field and length $B_s l$, while outside the solenoid, the magnetic field quickly reduces to a negligible level compared to the internal field, and so the integral of **B**. **dr** in this region may be approximated to zero. Hence,

$$\oint_C \mathbf{B} \cdot \mathbf{dr} = B_s l \tag{3.15}$$

We should note that there will clearly be a region at either end of the solenoid where the magnetic flux lines are just starting to spread out. In this region, the magnetic field is not negligible compared with the internal field, and so the result will only be valid for long solenoids where this effect can be reasonably ignored. We now need to calculate the total current passing through the Amperian loop. Every time the spiral of wire passes round the coil, it threads through the Amperian loop, and so the total current passing through the loop is nI_s. It might feel a little like 'cheating' to have the same current counted multiple times, but it should be remembered that there is a different group of electrons at each point along the wire where the coil passes though the Amperian loop, and it is moving charge that produces magnetic fields. Therefore, the greater the number of turns of the coil, the more moving charge passes through the loop. Hence, Eq. (3.6) becomes

$$B_s = \frac{\mu_0 n I_s}{l} \tag{3.16}$$

Therefore, the magnetic flux density inside the solenoid is proportional to both the current I_s and the number of turns of the coil per unit length, n/l.

In both of these examples, what may appear to be a complex problem to solve due to a distribution of moving charges along wires becomes relatively trivial using the elegant mathematics of the Ampère circuital law, which is simply expressing the basic phenomenon that magnetic flux lines circulate around moving charges, and judicious choice of geometry for the Amperian loop employed in the calculation.

3.5 The Biot–Savart Law in Free Space and Superposition

While the Ampère circuital law is extremely powerful and enables the magnetic field in many situations to be evaluated, there are some circumstances where it proves to be simpler to use a general expression for the magnetic field around a small length of a flow of current. The magnetic flux density at any point arising from an arbitrary flow of current around a circuit may be determined by dividing the circuit into many such small elements of current flow. The resulting magnetic field at a point may be evaluated by simply summing these by integration using the principle of superposition (see Section 1.6). This approach is analogous to using the simple expression for the electric field around a point charge to calculate the electric field due to a complex charge distribution as being the sum of the fields that would result from dividing the charge distribution into many point charges in electrostatics.

This 'general expression' is called the *Biot–Savart law*, and is given by

$$\mathbf{dB} = \frac{\mu_0 I}{4\pi} \frac{\mathbf{dl} \times \mathbf{r}}{|\mathbf{r}|^3} \tag{3.17}$$

where a current I flowing along a short length given by the vector $\mathbf{dl}$ produces a magnetic flux density $\mathbf{dB}$ at a position relative to the current given by the vector $\mathbf{r}$, as shown in Figure 3.5. The Biot–Savart law can then be integrated around a complete current loop to find the magnetic flux density at a point.

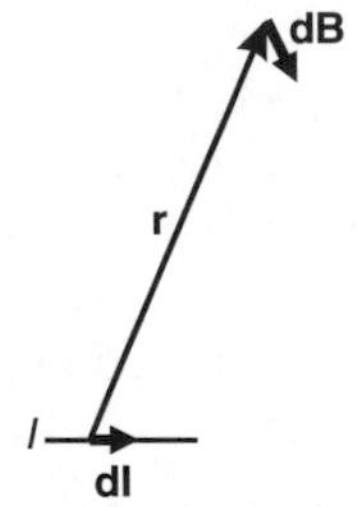

For example, let us find an expression for the magnetic field along an axis running through the centre of a circular loop of current I with a radius r_l in free space, as shown in Figure 3.6. The magnetic field will vary with distance x from the plane of the current loop. The Biot–Savart law can be applied to solve this problem by dividing the current loop up into small lengths $\mathbf{dl}$, where each length will subtend a small angle $d\phi$ around the loop such that

$$|\mathbf{dl}| = r_l d\phi \tag{3.18}$$

Figure 3.6 shows the direction of $\mathbf{dB}$ that results at a distance x along the axis away from the loop. It is clear that $\mathbf{dB}$ has a component that points along the axis, dB_x, and a radial component. However, from the cylindrical symmetry of the system, when the magnetic flux density components produced by all of the line elements around the circular loop are summed, the radial components of magnetic field will cancel out, and therefore only the axial component needs to be considered. By applying the Biot–Savart law (Eq. (3.17)) to the small element shown, the axial magnetic flux density can be expressed as

$$dB_x = \frac{\mu_0 I}{4\pi |\mathbf{r}|^3} |\mathbf{dl}||\mathbf{r}| \sin \theta \tag{3.19}$$

Figure 3.5 The Biot–Savart law allows the magnetic flux density $\mathbf{dB}$ to be calculated at a position $\mathbf{r}$ relative to a current I flowing along a short path length $\mathbf{dl}$.

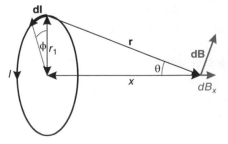

Figure 3.6 The magnetic flux density $\mathbf{dB}$ produced by a small length $\mathbf{dl}$ of a current loop of radius r_l and the axial component of the magnetic flux density dB_x.

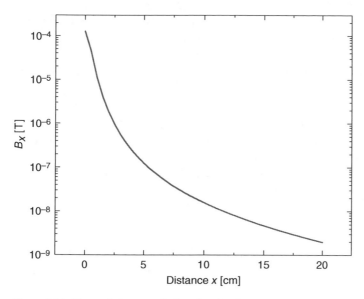

Figure 3.7 The radial magnetic flux density B_x produced by 1 A current loop of 1 cm diameter as a function of distance from the loop.

Therefore, using Pythagoras' theorem to express $|\mathbf{r}|$ in terms of x and r_l, and $|\mathbf{dl}| = r_l d\phi$, we have

$$dB_x = \frac{\mu_0 I \sin\theta}{4\pi \left(x^2 + r_l^2\right)} d\phi \tag{3.20}$$

We can also substitute that $\sin\theta = r/(x^2 + r^2)^{1/2}$ to give

$$dB_x = \frac{\mu_0 I r_l^2}{4\pi \left(x^2 + r_l^2\right)^{3/2}} d\phi$$

which must be integrated around the current loop from $\phi = 0$ to $\phi = 2\pi$ to evaluate the total axial magnetic field as a function of distance from the current loop:

$$B_x = \int_0^{2\pi} \frac{\mu_0 I r_l^2}{4\pi \left(x^2 + r_l^2\right)^{3/2}} d\phi = \frac{\mu_0 I r_l^2}{2\left(x^2 + r_l^2\right)^{3/2}} \tag{3.21}$$

Figure 3.7 shows how the radial magnetic flux density varies with distance from a 1 cm diameter loop of wire carrying a current of 1 A in free space. For large distances from the loop where $x \gg r_l$, the magnetic flux density decays as $\sim r_l^{-3}$, and so the field decreases very quickly for even relatively short distances from the loop.

3.6 The Gauss Law of Magnetic Fields in Free Space

In Section 1.5, we saw that the Gauss law of electrostatics expresses mathematically the fundamental physical effect that charge produces electric fields. It does this by considering the total number of flux lines passing through a closed Gaussian surface: lines which leave the surface count positively and those which enter count negatively. The total number of flux lines will only be non-zero if there is a net charge inside the Gaussian surface for flux lines to either begin or end on.

However, whereas lines of electric field begin and end on charges, the situation for magnetic fields is very different – instead the magnetic flux lines form closed loops around moving charges, as discussed in Section 3.2. Therefore, we could imagine a closed Gaussian surface at any point in free space. If we again consider the total number of flux lines passing through the surface, with those leaving counting positively and those entering counting negatively, we will always end with a sum of zero as any individual flux line which enters the surface must also leave as it has no beginning or end; it is a closed loop. Consequently, we can write a Gauss law of magnetic fields as

$$\oint_S \mathbf{B} \cdot \mathbf{dA} = 0 \tag{3.22}$$

where the left-hand side of the equation is the integral over a closed surface of the dot product of the magnetic field and local surface vectors $\mathbf{dA}$ (see Section 1.4 for a detailed discussion of the mathematics behind this expression).

Although the Gauss law of magnetic fields is rarely practically used by itself in this form, when we look at the Maxwell equations in Section 5.3, we shall see that this becomes a very important equation and one with considerable power in differential form.

3.7 The Faraday Law of Electromagnetic Induction in Free Space

The Lorentz force law (Eq. (3.3)) states that a charge which is moving in a magnetic field will experience a force. Therefore, if a solid object which contains mobile charges, such as a metal with its 'sea' of delocalized valence electrons, is moved through a magnetic field, then the mobile charges will move as a result of the Lorentz force which they experience.

By way of an example, let us consider a metal wire of length l moving at a velocity $\mathbf{v}$ through a uniform magnetic field $\mathbf{B}$ in free space, as shown in Figure 3.8. The Lorentz force is proportional to the cross product of the velocity of the charges and the magnetic field, and so by choosing the magnetic flux density, the alignment of the wire and the direction of motion of the wire to all be perpendicular to each other, the mathematics is simplified (although basic trigonometry allows other orientations to be calculated relatively trivially).

The motion of the wire in the magnetic flux density will cause the mobile charges to experience a force which acts along the axis of the wire, and so a charge distribution will form so that an electric field is produced in the wire. This electric field will cause a force to act on the mobile charges which opposes the magnetic force, and an equilibrium condition will be set up after a short time so that these two forces balance. It is clear from Eq. (3.4) that when the total force $\mathbf{F}$ acting on the charges is zero,

$$\mathbf{E} = -\mathbf{v} \times \mathbf{B} \tag{3.23}$$

Figure 3.8 An electric field **E** is set up in a wire of length *L* moving with a velocity **v** in a magnetic flux density **B**. This leads to an induced potential difference V_l between the ends of the wire.

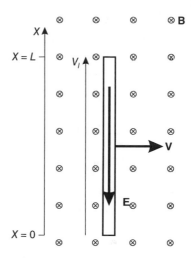

This electric field is uniform along the length of the wire, and so there must be a potential difference between the two ends of the wire which (from Eq. (1.9)) is

$$V_l = -\int_l^0 \mathbf{E} \cdot \mathbf{dx} = \int_l^0 (\mathbf{v} \times \mathbf{B}) \cdot \mathbf{dx} \tag{3.24}$$

Therefore, remembering that **v** and **B** are perpendicular, this reduces in this case to

$$V_l = |\mathbf{v}||\mathbf{B}|l \tag{3.25}$$

The motion of a metal wire in a magnetic field leads to a potential difference being induced between the two ends of the wire.

It is very rare that we would ever be in a practical situation where we had a simple length of wire inside a magnetic flux density. Instead, we would normally have a closed circuit around which current could flow. Figure 3.9 shows a simple extension of the situation in Figure 3.8 where we have a square loop of wire whose plane is perpendicular to the uniform magnetic flux density, and which is again moving at a velocity **v** which is perpendicular to the magnetic field. If the diameter of the wire is small compared with the length of the sides of the square loop, then the potential difference induced across the width of the sides of the wire loop that are in the same direction as the velocity will be negligible compared with the potential difference induced between the ends of the other two sides of the wire loop. However, as the potential difference across these two sides is pointing in the same physical direction, the total potential difference between the ends of the loop of wire will be zero – the

Figure 3.9 There is no potential difference induced between the ends of a square loop of wire moving at a velocity **v** in a uniform magnetic field **B**.

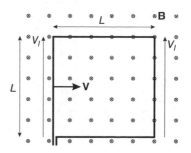

two voltages will cancel each other out. A potential difference between the two ends of the wire will only form if the two sides of the loop are in regions of different magnetic flux density **B**. As this variation in magnetic flux density is in the same direction as the velocity of the loop, we can conclude that this means that the total flux enclosed by the wire loop must be changing with time. This is expressed by the Faraday law of electromagnetic induction, which states that a potential difference is developed on a loop of wire given by

$$V = -\frac{d\Phi}{dt} \tag{3.26}$$

where Φ is the total magnetic flux passing through the wire and therefore $d\Phi/dt$ is the rate of change of flux in the loop.

It is possible to show that this is true by considering the moving loop shown in Figure 3.10 where the sides in the direction of the velocity **v** have a very small length δx. The magnetic flux density will be uniform in the y- and z-direction, but varies along the x-direction, and so can be written as **B**(x). The total potential difference developed around the loop is the sum of the potential differences on each of the four sides,

$$V = V_1 + V_2 + V_3 + V_4 \tag{3.27}$$

We already know that no significant potential difference is developed on the sides of the loop that are parallel to the velocity, so $V_2 = V_4 = 0$. Using the result for the potential difference developed across a single wire in a magnetic field (Eq. (3.25)), the potential difference developed across side 1 of the loop must be

$$V_1 = vB(x)l \tag{3.28}$$

where $v = |\mathbf{v}|$ and $B(x) = |\mathbf{B}(x)|$. Noting the direction of V_3 in Figure 3.10, the potential difference developed across side 3 of the loop is similarly

$$V_3 = -vB(x + \delta x)l \tag{3.29}$$

A simple first-order approximation allows $B(x + \delta x)$ to be expressed as

$$B(x + \delta x) \approx B(x) + \frac{\partial B}{\partial x}\delta x \tag{3.30}$$

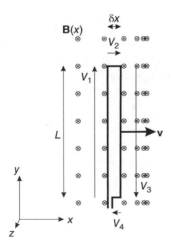

Figure 3.10 A rectangular loop of wire with side lengths L and δx moving with a velocity **v** through a spatially varying magnetic field **B**(x).

Substituting for all the potential differences of the individual sides of the loop into Eq. (3.27) gives

$$V = -vl\frac{\partial B}{\partial x}\delta x \tag{3.31}$$

which can be rearranged to give

$$V = -v\frac{\partial}{\partial x}(Bl\delta x) \tag{3.32}$$

The area of the loop is $l\delta x$, and therefore $Bl\delta x$ is the flux enclosed by the loop $\delta\Phi$. Meanwhile, the velocity can be expressed as $v = \partial x/\partial t$, so Eq. (3.32) reduces to the Faraday law

$$V = -\frac{\partial x}{\partial t}\frac{\partial \Phi}{\partial x} = -\frac{d\Phi}{dt} \tag{3.33}$$

In practice, any situation in which the flux enclosed by a loop of wire changes with time will lead to a potential difference across the loop. This is often called an *induced electromotive force* (e.m.f.). The movement of the loop through a spatially varying magnetic field, as we have already considered, is one clear example, but a time-varying magnetic field passing through a static loop will also have the same effect. The significance of the minus sign in Eq. (3.26) (which is technically the *Lenz law*) is that the direction of the induced e.m.f. is to try to drive a current through the loop and create a magnetic field which would oppose the change in flux. We shall consider a case where an induced e.m.f. is produced by a time-varying magnetic field in Section 3.8.

3.8   Inductors and Energy Storage

In Section 3.4, we calculated the magnetic flux density that exists inside a simple coil of wire carrying a current I, as shown in Figure 3.4. If the current is not a d.c. current but an a.c. current with an angular frequency ω, then this can be expressed as

$$I_s(t) = I_{s0}\cos(\omega t) \tag{3.34}$$

Therefore, it is clear from Eq. (3.16) that the magnetic flux density, and hence the total flux, inside the coil will be varying with time. From the Faraday law of electromagnetic induction (Eq. (3.26)), the result will be that a potential difference is developed across the ends of the coil.

We need to calculate the total flux inside the coil. We have assumed previously that the magnetic field is uniform inside the coil, which is a reasonable approximation for a long coil of relatively small radius R. It is clear from Figure 3.4 that the lines of magnetic flux run down the length of the coil, and so each turn of the coil encloses the flux lines. The flux in one cross section through the coil which would be enclosed by a single turn of the coil is

$$\phi = B_s\pi R^2 = \frac{\mu_0 n\pi R^2 I_s}{l} \tag{3.35}$$

where Eq. (3.16) has been used to substitute for B_s. The total flux enclosed by all n turns of the coil is then

$$\Phi = n\phi = \frac{\mu_0 n^2 \pi R^2 I_s}{l} \tag{3.36}$$

Between Eqs. (3.36) and (3.34) we have expressions for how the total flux enclosed by the coil depends on current, and how the current in the coil depends on time. Therefore, we can use the chain rule to similarly break up the Faraday law of electromagnetic induction (Eq. (3.26)) into

$$V = -\frac{d\Phi}{dI}\frac{dI}{dt} \tag{3.37}$$

The change in flux through a coil of wire produced by a change in current is called the *inductance* of the coil,

$$L = \frac{d\Phi}{dI} \tag{3.38}$$

and its unit is the henry (H). Substitution of Eq. (3.38) into Eq. (3.37) gives an explicit expression linking the potential difference across an inductor to the current through it:

$$V = -L\frac{dI}{dt} \tag{3.39}$$

The purpose of making such a definition is that inductance does not depend on how the current is actually varying with time in the coil as this is separated out. Hence, the inductance can be determined for a coil without needing to know about the form of the time-varying current. For the coil of Figure 3.4, differentiating Eq. (3.36) with respect to I_s gives

$$L = \frac{\mu_0 n^2 \pi R^2}{l} \tag{3.40}$$

The inductance is solely dependent on the geometry of the coil and the medium in which it exists (in this case free space). It is worth noting that the inductance is not linearly related to the number of turns on the coil but instead is proportional to n^2. This is a common feature of expressions for inductance, and it is a consequence of the fact that each additional turn of the coil not only leads to a greater magnetic field and therefore flux in one cross section through the coil, but also wraps around this flux once more, leading to a greater induced e.m.f.

It should be noted that the expression for inductance (Eq. (3.38)) is very similar in form to that for capacitance (Eq. (2.25)). The capacitor is an energy storage device (Section 2.3) as an electric field fills the volume of space between the conductors (see Section 1.3). Similarly, energy is stored in inductors as there is a magnetic field filling the volume within the coil, which is given by

$$U = \frac{1}{2}LI^2 \tag{3.41}$$

Reference

Einstein, A. (1905). Zur Elektrodynamik bewegter Körper. *Annalen der Physik* 322 (10): 891–921.

4

Magnetic Fields in Materials

4.1 The Interaction of Magnetic Fields with Matter

In Chapter 3, we confined ourselves to considering the properties of magnetic fields in free space. However, it is very common for magnetic fields to interact with matter which itself consists of atoms and, depending on its nature, molecules. In this chapter, we will consider this interaction in detail. We will begin by considering a simple case of an ideal linear magnetic material to give us some basic relationships between key quantities that are used to describe magnetism before considering some real magnetic materials, including diamagnetic materials, ferromagnetic materials and paramagnetic materials. We will conclude with a discussion on magnetic circuits and electromagnets.

4.2 The Magnetization of Ideal Linear Magnetic Materials

Electrons exist in well-defined orbitals around atoms, as determined by quantum mechanics, and these behave like small loops of current. Quantum mechanics also shows us that electrons themselves have an intrinsic angular momentum that is normally called *spin*, and which can be thought of as the electron spinning around on own axis. This will also produce a small loop of current. Therefore, each microscopic constituent of matter, whether individual atoms or molecules, can be considered to have a net current loop associated with it. We can define the effective current flowing around this loop to be I and we can also define an effective surface normal vector **A** whose magnitude is the area of the current loop. Its direction is perpendicular to the plane of the loop and pointing away when the current appears clockwise as shown in Figure 4.1 (this being consistent with the conventions for the Amperian loop considered in Section 3.3). From this we can ascribe a small magnetic dipole moment **m** also shown in Figure 4.1, with each of these microscopic constituents of matter,

$$\mathbf{m} = I\mathbf{A} \tag{4.1}$$

The unit of the magnetic dipole moment is $A\,m^2$. It is worth noting the strong similarity between this view of magnetic dipole moments at the microscopic level and that of electric dipole moments (see Section 1.7 and particularly Eq. (1.23)). These magnetic dipole moments are effectively microscopic versions of the current loop that we considered in

Electromagnetism for Engineers, First Edition. Andrew J. Flewitt.
© 2023 John Wiley & Sons Ltd. Published 2023 by John Wiley & Sons Ltd.
Companion website: www.wiley.com/go/flewitt/electromagnetism

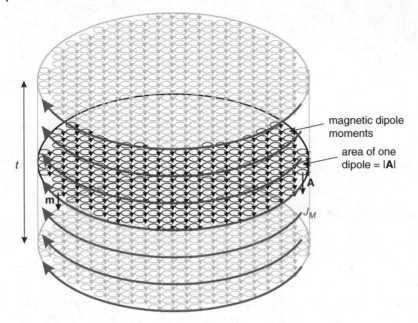

t

m

magnetic dipole moments

area of one dipole = |**A**|

A

J_M

Figure 4.1 A magnetic material containing many small dipole moments *m*, each of area *A*. These have the same overall effect as a surface magnetization current J_M.

Section 3.5. By inspection, we can rewrite Eq. (3.21), which expresses the axial component of the magnetic flux density as a function of axial distance *x* from the loop, in terms of the magnetic dipole moment as

$$B_x = \frac{\mu_0 |\mathbf{m}|}{2\pi x^3} \tag{4.2}$$

for the situation where *x* is much greater than the radius of the effective current loop.

If an external magnetic flux density **B** is applied to a material, for example by placing a rod of a particular material inside a solenoid (see Section 3.4), then each of the magnetic dipole moments will experience a torque

$$\mathbf{T} = \mathbf{m} \times \mathbf{B} \tag{4.3}$$

which will tend to cause them to align with the magnetic flux so that the torque reduces towards zero. If there are a number density of *N* magnetic dipole moments per unit volume in the material, and the average component of their individual magnetic dipole moments that aligns in the direction of the applied magnetic flux density is $\mathbf{m_B}$, then we can define a *magnetization* **M** of the material due to the applied field as the aligned dipole moment per unit volume, which will be given by

$$\mathbf{M} = N\mathbf{m_B} \tag{4.4}$$

The magnetization is effectively a total magnetic dipole moment per unit volume, as it is the sum of all the individual dipole moments in a unit volume of the material. The unit of the magnetization is therefore that of the dipole moment ($A\,m^2$) divided by volume (m^3), which gives $A\,m^{-1}$.

This is analogous to the electric polarization seen in dielectrics that are subjected to an externally applied electric field (see Section 2.2). In the electric case, we saw that the polarization could be considered to be effectively due to the presence of a surface charge density. Similarly, we can consider the magnetization to be due to a current running over the surface of the material. Figure 4.1 shows many aligned magnetic dipole moments in a plane through a material. In the bulk of the material, the local current loops all cancel each other out. However, on the surface there is a net current which appears to form a loop over the entire surface, the effect of which yields the observed magnetization. No single electron moves around the entire current loop – they are all moving in individual local loops in reality – but the effect is the same.

This effective surface current due to magnetization J_M scales with the thickness of the material, as indicated by the dimension t in Figure 4.1. Therefore, it has units of A m^{-1}, as the total surface magnetization current flowing would be $J_M t$. In terms of this current, the total dipole moment $\mathbf{m}_t$ of the small volume of material shown in Figure 4.1 of area $\mathbf{A}_t$ and thickness t would therefore be

$$\mathbf{m}_t = J_M t \mathbf{A}_t \tag{4.5}$$

However, knowing that magnetization $\mathbf{M}$ is the magnetic dipole moment per unit volume allows us to also express the magnitude of $\mathbf{m}_t$ as

$$m_t = M t |\mathbf{A}_t| \tag{4.6}$$

Therefore, equating Eqs. (4.5) and (4.6) shows that

$$J_M = |\mathbf{M}| \tag{4.7}$$

or in other words, the magnetization is the effective surface current whose effect would be to produce the observed magnetic dipole moment of the material.

Let us assume that we have a material in which all of the microscopic magnetic dipole moments are randomly aligned when no external field is applied. Inspection of the expressions for the magnetic flux density produced by a dipole (Eq. (4.2)) and the magnetization due to the dipoles (Eq. (4.3)) would lead us to expect that the magnetization induced is proportional to $\mathbf{B}/\mu_0$. The constant of proportionality χ_B is called the *magnetic susceptibility*, so that

$$\mathbf{M} = \chi_B \frac{\mathbf{B}}{\mu_0} \tag{4.8}$$

It is analogous to the electric susceptibility (Eq. (2.9)) and is also similarly dimensionless.

We now introduce a new quantity: the magnetic field $\mathbf{H}$. This is analogous to the electric field $\mathbf{E}$, and, like the electric field, it is a measure of the actual field that will produce a real force acting on a moving charge in the presence of the field. The magnetic flux density $\mathbf{B}$ that we have been using throughout Chapter 3 is like the electric flux density $\mathbf{D}$; it is simply a mathematical construction which makes calculation of magnetic fields more straightforward.

For the simple magnetic material which we have been considering where the linear relation of Eq. (4.8) between magnetization and applied magnetic flux density is valid, a simple linear relation also holds between $\mathbf{B}$ and $\mathbf{H}$ given by

$$\mathbf{B} = \mu_0 \mu_r \mathbf{H} \tag{4.9}$$

which is analogous to the relationship between electric field and electric flux density (Eq. (2.15)). We have already defined μ_0 as being the permeability of free space (see Section 3.3), while μ_r is known as the *relative permeability* and is a material-dependent quantity. The relative permeability of free space is unity. Together, the quantity $\mu_0\mu_r$ is known simply as *permeability*.

We will now mathematically link these various magnetic quantities together in a similar fashion to Eqs. (2.9) and (2.13) for the polarization of dielectrics by once again considering our magnetic material to be inside a solenoid, as shown in Figure 4.2. In Section 3.4, we were able to use the Ampère circuital law in free space (Eq. (3.6)) to calculate the magnetic flux density inside the empty solenoid. We can again apply Eq. (3.6) to this scenario by using the Amperian loop shown as the dashed line in Figure 4.2. However, the current flowing through the Amperian loop now has two components: the current in the coil of the solenoid, $I = nI_s$, and the effective surface magnetization current, which we will call I_M. Remembering that **M** is the surface current per unit length, we can calculate I_M by integrating **M** around the Amperian loop according to

$$I_M = \oint_C \mathbf{M}.\mathbf{dr} \tag{4.10}$$

As the magnetization in free space is zero, this will effectively return the component of magnetization along the line of the Amperian loop multiplied by the length of the magnetic material, as required to give the surface magnetization current. Equation (3.6) therefore becomes

$$\oint_C \mathbf{B}.\mathbf{dr} = \mu_0(I + I_M) \tag{4.11}$$

$$\oint_C \mathbf{B}.\mathbf{dr} = \mu_0 I + \mu_0 \oint_C \mathbf{M}.\mathbf{dr} \tag{4.12}$$

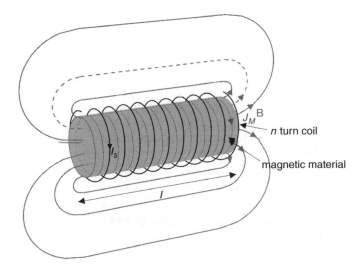

Figure 4.2 A magnetic material with a surface magnetization current J_M inside a solenoid with a current I_S.

This can be rearranged to give

$$\oint_C \left(\frac{\mathbf{B}}{\mu_0} - \mathbf{M} \right) . \mathbf{dr} = I \tag{4.13}$$

This then leads to the general definition of the magnetic field **H** which applies to both linear and nonlinear magnetic materials as

$$\mathbf{H} = \frac{\mathbf{B}}{\mu_0} - \mathbf{M} \tag{4.14}$$

which relates the magnetic field, magnetic flux density and magnetization, just as Eq. (2.13) related the electric field, electric flux density and polarization in a dielectric. The Ampère circuital law may also be re-expressed by substituting Eq. (4.14) into Eq. (4.13),

$$\oint_C \mathbf{H} . \mathbf{dr} = I \tag{4.15}$$

which can be used for all magnetic materials, whereas the form used in Eq. (3.6) can only be used in free space.

Substitution of Eq. (4.8) into Eq. (4.14) allows the magnetic field and flux density to be related directly by susceptibility for linear magnetic materials,

$$\mathbf{B} = \frac{\mu_0 \mathbf{H}}{1 - \chi_B} \tag{4.16}$$

and comparison of this equation with Eq. (4.9) reveals the link between relative permeability and magnetic susceptibility to be

$$\mu_r = \frac{1}{1 - \chi_B} \tag{4.17}$$

Whether a magnetic material can be considered to be linear or not has a significant bearing on which of the above equations are valid and therefore on the approach that can be taken to determining how the material will behave in particular applications. In the following sections, we will consider a number of different classes of magnetic materials, how they perform, and under what circumstances they may be considered to be linear or otherwise.

4.3 Diamagnetic Materials

We know intuitively that many elements appear not to be magnetic in nature, with a good example of this being carbon. In fact, carbon does respond to magnetic fields, albeit weakly, and is one of the class of *diamagnetic* materials, with other elemental examples being boron, silicon, sulphur, copper, and gold.

In Section 4.2, we reminded ourselves that the electrons that surround atoms exist in quantum mechanically defined orbitals with some associated angular momentum **L**. If there are Z electrons around a particular atom, then the magnetic moment for any one of these electrons $\mathbf{m}_n$ will be

$$\mathbf{m}_n = \frac{-e}{2m_e} \mathbf{L}_n \tag{4.18}$$

where m_e is the free mass of the electron. This is consistent with Eq. (4.1), as the magnitude of the angular momentum is determined by the electron velocity in that state v_n and the effective orbital radius of that state r_n according to

$$|\mathbf{L}_n| = m_e v_n r_n \tag{4.19}$$

The current I_n associated with the electron in that state is

$$I_n = \frac{-ev_n}{2\pi r_n} \tag{4.20}$$

remembering the negative charge on the electron, and the effective area of the orbital is πr_n^2. Therefore, multiplying the effective area and current and substituting for the angular momentum in Eq. (4.19) gives Eq. (4.18). The total magnetic moment associated with any one atom is then the sum of that due to all of the n electrons, and so is

$$\sum_{n=1}^{Z} \mathbf{m}_n = \frac{-e}{2m_e} \sum_{n=1}^{Z} \mathbf{L}_n \tag{4.21}$$

The sum of the angular momentum of all the electrons in diamagnetic materials is zero, and so they have no net magnetic dipole of their own. It is for this reason that they appear not to be magnetic in nature.

Let us imagine that an external magnetic flux density $\mathbf{B}$ is now applied to the region of space in which the atom exists and that a time t is taken to increase the flux density from 0 to $\mathbf{B}$. If we consider each electron to be a current loop, it is clear that each experiences a changing flux inside its loop where the rate of change of flux during the initial period Δt is

$$\frac{d\Phi}{dt} = \frac{\mathbf{B}.\mathbf{A}_n}{\Delta t} \tag{4.22}$$

where $\mathbf{A}_n$ is the vector area of the electron orbital, and so $\mathbf{B}.\mathbf{A}_n$ is the total flux passing through this area. As we saw in Section 4.2, the spins will tend to align with the applied magnetic field. The dot product between flux and vector area will then become a simple product and Eq. (4.22) becomes

$$\frac{d\Phi}{dt} = \frac{B\pi r_n^2}{\Delta t} \tag{4.23}$$

where $B = |\mathbf{B}|$. According to the Faraday law, an electromotive force will be induced whose action on the electron will be to change its angular velocity by $\Delta\omega_n$, and therefore its angular momentum. From Eqs. (3.26) and (4.23) we know that the induced e.m.f. round the whole current loop is

$$V_{\text{loop}} = -\frac{d\Phi}{dt} = -\frac{B\pi r_n^2}{\Delta t} \tag{4.24}$$

If the initial angular velocity of the electron is ω_n then the fraction of the whole current loop that the electron travels round in time Δt is $\omega_n \Delta t / 2\pi$, and so the actual potential difference experienced by the electron is

$$\Delta V = V_{\text{loop}} \frac{\omega_n \Delta t}{2\pi} = -\frac{B\omega_n r_n^2}{2} \tag{4.25}$$

This potential difference results in a change in kinetic energy

$$\Delta\left(\frac{m_e r_n^2 \omega_n^2}{2}\right) = -e\Delta V \tag{4.26}$$

If the radius of the orbit is unchanged, then the change in kinetic energy results only in a change in angular frequency, and Eq. (4.26) becomes

$$m_e r_n^2 \omega_n \Delta\omega_n = \frac{eB\omega_n r_n^2}{2}$$

$$\Delta\omega_n = \frac{eB}{2m_e} \tag{4.27}$$

There is therefore a change in the angular momentum of each electron in the direction of the magnetic flux density, which (remembering that $\Delta L = mr^2 \Delta\omega$) is

$$\Delta \mathbf{L}_n = m_e r_n^2 \frac{e\mathbf{B}}{2m_e} = \frac{r_n^2 e\mathbf{B}}{2} \tag{4.28}$$

From Eq. (4.18), this means that the magnetic moment of each electron in each atom also changes by

$$\Delta \mathbf{m}_n = -\frac{r_n^2 e^2 \mathbf{B}}{4m_e} \tag{4.29}$$

Therefore, for the whole atom with its Z electrons there will now be a total magnetic moment which can be calculated by determining the average contribution of any one of these electrons to be

$$\mathbf{m} = -\frac{Zr_0^2 e^2 \mathbf{B}}{6m_e} \tag{4.30}$$

where r_0 is the average radius of the electron orbits and where, for a spherically symmetric system, $\sum r_0^2 = \frac{2}{3}\sum r_n^2$. For a material with an atomic density of N, the magnetization induced due to the applied magnetic flux density is

$$\mathbf{M} = -\frac{NZr_0^2 e^2 \mathbf{B}}{6m_e} \tag{4.31}$$

The minus sign in this expression is very significant, as it shows that the induced magnetization in the diamagnetic material *opposes* the applied magnetic flux density. The magnetic field inside a diamagnetic material is actually less than that in free space for a given magnetic flux density. The magnetic susceptibility must therefore be negative for these materials, and this is made clear by comparing Eqs. (4.31) and (4.8) to give

$$\chi_B = -\frac{NZr_0^2 e^2 \mu_0}{6m_e} \tag{4.32}$$

As a consequence, diamagnetic materials will experience a force in an inhomogeneous magnetic flux density that causes them to move in the direction of reducing magnetic flux density.

The magnetic susceptibility of diamagnetic materials is extremely small. For example, it is -9.36×10^{-6} for copper. In practice, all materials show some diamagnetic behaviour, but it is so small compared with the other magnetic effects (which we will consider later in this chapter) that it is only of significance in materials where this is the only magnetic effect present, which tends to be those where all electrons are paired.

In 1933, it was discovered that superconductors (materials with no electrical resistance) expel magnetic flux from their interior (Meissner and Ochsenfeld 1933). This is known as the *Meissner effect*, and it is true for many superconductors up to some critical magnetic flux density. Although it is a very different mechanism by which the magnetic field inside the superconductor is reduced compared with that described for simple diamagnetic materials, superconductors are often thought of as having diamagnetic characteristics as the magnetic susceptibility is negative. In fact, the magnetic susceptibility of superconductors showing the Meissner effect is –1 (or very close to –1). As diamagnetic materials will move to regions of lower magnetic flux density, the diamagnetic effect in superconductors is so strong that it is possible to make superconducting magnets levitate against gravity in a spatially varying magnetic field.

4.4 Paramagnetic Materials

In Section 4.3, we saw that for a material to be diamagnetic, its constituent atoms or molecules must have zero total magnetic moments, as given by Eq. (4.21). However, in many cases, such as elements where there are unpaired electrons, the constituents will have a permanent magnetic moment. It might be expected that these individual magnetic moments would interact with each other to point in the same direction to minimize energy. However, if they are well separated, then the interaction is very weak compared with the thermal energy available which acts to disorder the magnetic moments, and as a result, the dipoles only align to an externally applied magnetic flux density. These *paramagnetic* materials are characterized by having no net magnetic field of their own in the absence of an applied magnetic flux. Examples of paramagnetic elements include sodium, magnesium, aluminium, titanium, manganese and tin. The paramagnetic effect is much larger than the diamagnetic effect, which is insignificant in these materials.

In order to derive an equation for the magnetic susceptibility in paramagnetic materials, we shall consider a uniform material which has a number density N of constituent magnetic moments $\mathbf{m}$. In the case of an elemental material, such as aluminium, N will be the number density of atoms. In the presence of an applied magnetic flux density $\mathbf{B}$, it is energetically favourable for the magnetic moments to align in the same direction as the flux density. We can define a potential energy associated with a magnetic moment at some arbitrary angle θ with respect to the direction of $\mathbf{B}$ to be $-\mathbf{m}.\mathbf{B}$, where the zero reference potential energy is for a magnetic moment that is perpendicular to the magnetic flux density. In this way, magnetic moments that are aligned with the flux have a negative potential energy and those that are aligned against the flux have a positive potential energy with respect to zero. As there is no restriction on the number of magnetic moments that can point in a given direction, the system will obey Boltzmann statistics, and so the number of moments aligned at some particular angle θ will be

$$n(\theta)d\theta = n_0 \exp\left(\frac{-\mathbf{m}.\mathbf{B}}{k_B T}\right) 2\pi \sin\theta \, d\theta \tag{4.33}$$

where n_0 is a pre-exponential constant, and $2\pi \sin \theta \, d\theta$ is the solid angle subtended by the small angle $d\theta$ at some angle θ. For paramagnetic materials where $\mathbf{m}.\mathbf{B} \ll k_B T$, a first-order Taylor approximation of the exponential term can be applied, simplifying Eq. (4.33) to

$$n(\theta)d\theta = n_0 \left[1 + \frac{\mathbf{m}.\mathbf{B}}{k_B T} \right] 2\pi \sin \theta \, d\theta$$

and expressing $\mathbf{m}.\mathbf{B}$ explicitly in terms of θ gives

$$n(\theta)d\theta = n_0 \left[1 + \frac{mB \cos \theta}{k_B T} \right] 2\pi \sin \theta \, d\theta \tag{4.34}$$

where $m = |\mathbf{m}|$ and $B = |\mathbf{B}|$. The constant n_0 can then be found by integrating $n(\theta)d\theta$ over all angles, as this must equal the total number of magnetic moments per unit volume:

$$N = \int_0^\pi n(\theta)d\theta = \int_0^\pi n_0 \left[1 + \frac{mB \cos \theta}{k_B T} \right] 2\pi \sin \theta \, d\theta = 4\pi n_0$$

Hence, $n_0 = N/4\pi$ and Eq. (4.34) becomes

$$n(\theta)d\theta = \frac{N \sin \theta}{2} \left[1 + \frac{mB \cos \theta}{k_B T} \right] d\theta \tag{4.35}$$

Finally, the magnetization can be found by integrating the components of all of the magnetic moments pointing in the direction of the magnetic flux density,

$$\mathbf{M} = \hat{\mathbf{B}} \int_0^\pi n(\theta)m \cos \theta \, d\theta \tag{4.36}$$

where $\hat{\mathbf{B}}$ is the unit vector pointing in the direction of the magnetic flux density. This can be evaluated as

$$\mathbf{M} = \hat{\mathbf{B}} \int_0^\pi \frac{N \sin \theta}{2} \left[1 + \frac{mB \cos \theta}{k_B T} \right] m \cos \theta \, d\theta$$

$$= \frac{Nm^2 \mathbf{B}}{3k_B T} \tag{4.37}$$

It is clear that, unlike in diamagnetic materials, the magnetization in paramagnetic materials is in the same direction as the magnetic flux density. Therefore, the magnetic susceptibility in paramagnetic materials is positive, as can be seen by comparing Eq. (4.37) with Eq. (4.8) to give

$$\chi_B = \frac{N\mu_0 m^2}{3k_B T} \tag{4.38}$$

Consequently, paramagnetic materials will experience a force in an inhomogeneous magnetic field that causes them to move in the direction of increasing magnetic flux density.

The magnetic susceptibility of paramagnetic materials is larger than for diamagnetic materials, but is still small compared with ferromagnetic materials (see Section 4.5). For example, the magnetic susceptibility of aluminium is 2.2×10^{-5}.

Another key difference between the magnetic susceptibility of diamagnetic and paramagnetic materials that is clear by comparing Eqs. (4.38) and (4.32) is that the former shows no temperature dependence whereas the latter does show a temperature dependence of the form

$$\chi_B = \frac{C}{T} \tag{4.39}$$

This effect was first observed by Pierre Curie in paramagnetic gases, and so is called the *Curie law*, with C being the Curie constant (Curie 1895).

4.5 Ferromagnetic Materials

We have considered paramagnetic materials in Section 4.4 where the constituent atoms or molecules in a solid do have a permanent magnetic dipole moment as a result of unpaired electrons, but where these dipoles do not interact, so there is no spontaneous magnetization. However, in some materials, and particularly those with extensive 3*d* electrons, there is a significant interaction. In *ferromagnetic* materials, it is energetically favourable for the neighbouring magnetic dipole moments to align parallel to each other due to a quantum mechanical interaction, even in the absence of an externally applied magnetic flux density. The most common element which displays this phenomenon is iron, from where ferromagnetism gets its name, but cobalt and nickel are other elemental examples.

The energy gain associated with this alignment of magnetic moments is not isotropic, but is greatest when the magnetization aligns with particular crystallographic directions. For example, alignment of the magnetization along the [100] direction in iron entails a significantly lower energy than alignment along the [111] direction. Such favoured magnetization alignments are known as *easy directions*. Most materials are not composed of a single crystal, but have a polycrystalline structure of grains with different crystallographic orientations, so different regions in a ferromagnetic material will have different magnetization directions, as shown in Figure 4.3. Local volumes with a uniform alignment are known as *domains* and are separated by *domain walls*. The width of the domain walls is determined by two competing energetic requirements. Within the domain wall, there will be a transition of magnetic moments pointing in the direction of the magnetization on one side of the wall to that on the other. Neighbouring magnetic moments will therefore be misaligned by a small angle, which requires energy. The wider a domain wall is, the smaller the angular difference will be between neighbouring magnetic moments within the wall. This would suggest that a wide domain wall would minimize energy. However, it is also the case that the magnetic moments within the wall will not be aligned with the easy direction, and this also requires energy. This would suggest that energy is minimized by having a narrow domain wall where the number of magnetic moments that are not aligned along the easy direction is minimized. In practice, it is the balance between these two competing energetic requirements that defines the width of the domain walls.

domain wall   domain

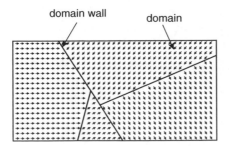

Figure 4.3 A schematic diagram showing the magnetic moments in a ferromagnetic material with well-defined domains within which all the magnetic moments are aligned. Domains are separated by domain walls. There is misalignment of the direction of the magnetic moments between different domains.

Figure 4.4 (a) An example of an arrangement of magnetic domains in a ferromagnetic material resulting in the magnetic flux density being confined within the material. (b) On application of an external magnetic field, the magnetization in some domains is rotated to align with the applied field and other domains are reduced in size where this is more energetically favourable than rotation of magnetization.

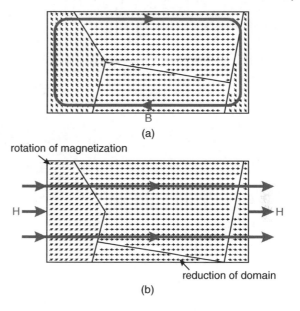

(a)

(b)

Although each domain in a sample of ferromagnetic material may have a magnetization, it is possible (and indeed often energetically favourable) for these to be aligned in such a way that all the flux lines are kept within the sample, so that there is no external flux density. An example of such an arrangement is shown in Figure 4.4a. If an external magnetic field is applied to a ferromagnetic material, then this will cause two effects shown in Figure 4.4b. One is a rotation of the magnetization within each domain to align with the direction of the applied field. The other is a movement of domain walls so that those domains whose magnetization naturally aligns closely with the applied magnetic field are enlarged at the expense of those where there is a large angular difference between the natural magnetization direction and the applied field. The latter process occurs where it would take a larger energy to rotate the magnetization within the existing domain than to move the domain wall.

These processes are highly nonlinear, and therefore the simple linear relationship between magnetic flux density and magnetic field of Eq. (4.9) no longer applies. In particular, once all highly misaligned domains have been eliminated at the expense of aligning well-aligned domains and these well-aligned domains have themselves had their magnetization rotated to align with the applied magnetic field, the system becomes saturated. No more magnetic flux lines can pass through the material, and so **B** cannot exceed a maximum value, even if the magnetic field **H** passing through the material is increased. If the applied magnetic field is reduced to zero, the ferromagnetic material will retain a net magnetization as this will tend to be energetically favourable. The magnetic flux density in the material in this condition is known as the *remnance flux*. A reverse magnetic field, known as the *coercivity*, must be applied in order to demagnetize the ferromagnetic material and return the internal magnetic flux density to zero. Graphs showing the relation between the magnitudes of **B** and **H** as well as these special conditions are plotted in Figure 4.5 for each of two different types of ferromagnetic materials: *soft ferromagnets* and *hard ferromagnets*.

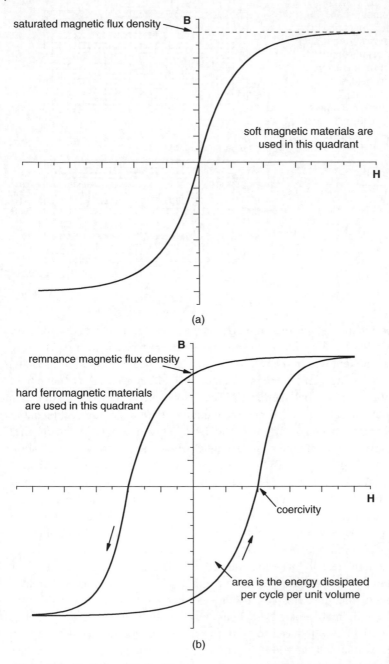

Figure 4.5 Typical magnetization curves for (a) a soft ferromagnetic material and (b) a hard ferromagnetic material.

In soft ferromagnetic materials, the coercivity required to demagnetize the material is low. Soft magnetic materials tend to be used as the core in transformers, which will be discussed in Chapter 12. These materials are often used in the top right-hand quadrant of the **B** versus **H** curve shown in Figure 4.5a, as the material is being used to guide a magnetic flux that is produced by one coil of wire on the primary side of the transformer so that it passes through the secondary coil as well. It should be easy to magnetize and demagnetize the material as the alternating current needs to keep reversing the direction of the magnetic flux. Hard ferromagnetic materials, however, have a high coercivity. They are very hard to demagnetize due to their microstructure and are commonly used as permanent magnets. Usually, we will want the permanent magnet to do mechanical work, and therefore they are often used in the top left-hand quadrant of the **B** versus **H** curve shown in Figure 4.5b so that work is being done against an applied external magnetic field.

4.6   Energy Stored in Magnetic Materials

In Section 1.3, we saw that energy is stored whenever a field occupies a volume of space. The consequence of this for magnetic fields in free space, which are created by currents, is that we can express the energy in a magnetic field in terms of inductance using Eq. (3.40). Likewise, energy must be stored when magnetic materials are magnetized.

In general, the energy stored in a magnetic field can be calculated as

$$U = \int_v \int_0^H \mathbf{B} . \, d\mathbf{H} \, dv \tag{4.40}$$

where v is the volume of space in which the field exists. This is analogous to Eq. (2.27) for the energy stored in an electric field. We will require the Maxwell equations to justify this expression (see Section 5.6), but for now we can consider **B.H** to be the component of magnetic flux density produced in the direction of an applied magnetic field multiplied by the magnitude of the magnetic field.

For the case of ideal linear magnetic materials of the type discussed in Section 4.2, where **B** and **H** can be simply related by permeability, we can substitute Eq. (4.9) into Eq. (4.40) to give

$$U = \int_v \int_0^B \frac{B}{\mu_0 \mu_r} \, dB \, dv = \int_v \frac{B^2}{2\mu_0 \mu_r} \, dv \tag{4.41}$$

This equation allows us to gain an insight into why the magnetic flux lines spread out around a solenoid, as shown in Figure 3.4. We know that lines of magnetic flux must form closed loops, and so the flux lines that are confined to run through the centre of the solenoid due to the current in the coil must eventually loop back on themselves. The B^2 term in Eq. (4.41) tells us that it costs a lot of energy to have a high flux density, and that halving the flux density will reduce the energy per unit volume by a factor of 4. Therefore, it is energetically favourable for the flux lines to quickly spread out at either end of the solenoid where the flux density is highest, as this leads to a significant reduction in the total energy of the system, even though the flux lines are occupying a bigger volume of space. However, as the flux lines continue to spread out, there comes a point where the reduction in total energy due to the reduced magnetic flux density is not as great as the increase in energy due to the greater volume that they occupy. The actual spread of the flux lines around the solenoid therefore represents the lowest total energy state of the system.

It is also clear, that it represents a lower energy per unit volume for a magnetic field to exist inside a material with a high relative permeability than a material with a low relative permeability such as air. Therefore, it is possible to guide magnetic flux lines around specific paths using loops of materials with a high μ_r, or to concentrate flux lines in a particular region using a block of high-permeability material. An example of the latter can be found in some AM or FM radio receivers where a cylinder of material with a high μ_r is inserted inside a coil of wire so that the magnetic field in the broadcast electromagnetic radio wave is concentrated inside the coil. The oscillating magnetic field induces an a.c. voltage in the coil which, with the aid of a resonant circuit to select a narrow range of frequencies associated with a given signal and a combination of amplifiers and demodulators, can be converted into sound.

The situation for the nonlinear materials is complicated by the lack of a simple expression that relates **B** and **H**. Instead, for soft ferromagnetic materials, a graphical integration of the area under a magnetization curve, such as in Figure 4.5a, has to be employed for evaluating Eq. (4.40). For hard ferromagnetic materials whose magnetization curves are of the form shown in Figure 4.5b, the area enclosed by the hysteresis of the curve is equal to the energy dissipated per cycle as a result of having to overcome the interaction between local magnetic dipole moments to make them all change direction. Ultimately this energy will be dissipated as heat. Therefore, soft ferromagnetic materials should be used for the core of a.c. electrical power transformers where the core is guiding an oscillating magnetic field (typically at 50 or 60 Hz) between two coils of wire to either step up or step down the voltage (see Section 12.3). Ideally, transformers should not consume power themselves, but allow efficient transfer of power to the desired loads on an electrical power grid.

4.7 The Magnetic Circuit

We have seen in Section 4.6 that it is energetically favourable for magnetic flux lines to exist in materials where a small magnetic field can produce a high magnetic flux density, such as is the case for linear magnetic materials with a high μ_r (see Section 4.2). Therefore, we can guide flux lines around specific loops, and an example of this is shown in Figure 4.6 where a small coil of wire with n turns and carrying a current I is wrapped around a short section of a toroid of magnetic material. If the cross-sectional radius r of the toroid is much smaller than the toroid radius R, then we can assume that the magnitude of the magnetic field H inside the toroid is uniform, and that no magnetic field leaks out of the magnetic material into the surrounding air. We can then apply the Ampère circuital law as expressed in Eq. (4.15) to an Amperian loop running through the centre of the toroid. In this case, **H** is parallel to **dr** at all points around the loop, which encloses all n turns of the coil, so Eq. (4.15) becomes

$$H \cdot 2\pi R = nI$$

$$H = \frac{nI}{2\pi R} \tag{4.42}$$

Figure 4.6 A coil of wire with n turns is wrapped around one part of a toroid of a linear magnetic material with a high μ_r such that the magnetic field is guided around a magnetic circuit within the material.

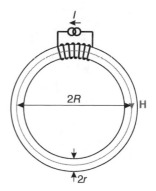

If the magnetic material has a well-defined relative permeability μ_r, then from Eq. (4.9) this becomes

$$B = \frac{\mu_0 \mu_r n I}{2\pi R} \tag{4.43}$$

and the total flux in the cross section of area $A = \pi r^2$ through any point in the toroid is

$$\phi = BA = \frac{\mu_0 \mu_r n I A}{2\pi R} \tag{4.44}$$

We can think of this as though the n turns of current I are producing the flux ϕ that circulates around the toroid in rather the same way that a voltage source can produce a current that circulates around a loop of wire. Such a system is often described as being a *magnetic circuit*, drawing an analogy with the electrical circuit. The quantity nI is then called the *magnetomotive force* (whereas we say in Section 3.7 that the electrical equivalent is the electromotive force V). By rearranging Eq. (4.44) to make the magnetomotive force the subject, we get

$$nI = \frac{2\pi R}{\mu_0 \mu_r A} \phi \tag{4.45}$$

This has a similar form to the Ohm law except that the proportionality between nI and ϕ is called the *reluctance* (whereas the proportionality between voltage and current in the Ohm law is resistance). In this case, the reluctance is

$$\mathcal{R} = \frac{2\pi R}{\mu_0 \mu_r A} \tag{4.46}$$

but in general, the reluctance of a loop of flux in a magnetic circuit is

$$\mathcal{R} = \frac{l}{\mu_0 \mu_r A}$$

where l is the length of the flux loop and A is the cross-sectional area. Therefore, if we have the scenario shown in Figure 4.7 of a coil of wire producing a flux which splits into two paths, we can sum the reciprocal of the reluctances of the two paths $\mathcal{R}_1$ and $\mathcal{R}_2$ to give an effective total reluctance $\mathcal{R}$ according to

$$\frac{1}{\mathcal{R}} = \frac{1}{\mathcal{R}_1} + \frac{1}{\mathcal{R}_2} \tag{4.47}$$

in much the same way that we would sum resistances in parallel.

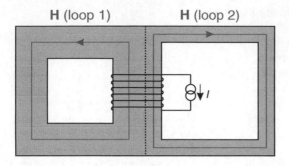

H (loop 1)   H (loop 2)

Figure 4.7 A coil of wire wrapped around the centre of a magnetic material with two well-defined magnetic circuits (loops 1 and 2) in parallel with each other. A reluctance can be assigned to each of the loops individually, $\mathcal{R}_1$ and $\mathcal{R}_2$. The total reluctance of the system is then given by Eq. (4.47).

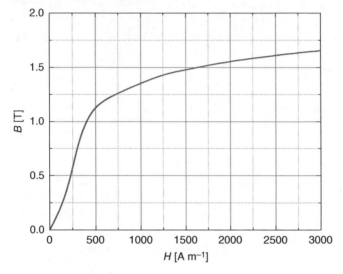

Figure 4.8 The graph showing the relation between the magnetic field H and the magnetic flux density B for cast steel, which is a soft ferromagnetic material.

The concept of the magnetic circuit allows us to quantitatively design magnetic systems. For example, if we wanted to generate a particular magnetic flux density in the toroid of Figure 4.6, then we could do this using Eq. (4.43), if we knew the relative permeability of the magnetic material. We could handle a similar situation for a nonlinear magnetic material if the relationship between B and H is known. Figure 4.8 is a graph showing the relationship between these quantities for cast steel, which is a soft ferromagnetic material. In this case, if we wanted to generate a magnetic flux density of 1.5 T inside the toroid, then this equates to a magnetic field of 1650 A m^{-1}. Equation (4.42) could then be used to determine the required magnetomotive force nI to produce this.

A particularly common problem is one where a magnetic material is being used to generate a particular magnetic flux density in a small volume of free space. This could be produced by having a small air gap of length l_g in a toroid of cross-sectional radius r and toroidal radius R, as shown in Figure 4.9. In this case, let us imagine that we are using a hard ferromagnetic material, such as 35% cobalt steel, as a permanent magnet, so that no current-carrying coil is required. Hard ferromagnetic materials are clearly nonlinear. Therefore, one approach to solving such a problem is to use a load line, where an equation

Figure 4.9 A toroid of a hard ferromagnetic material that is permanently magnetized produces a magnetic field in a small air gap of length l_g.

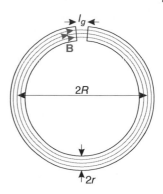

is generated which relates B and H in the magnetic material given that there is a linear relation between these two quantities in the air gap – this is called the *load line*. The load line is then plotted on top of the nonlinear characteristics of the magnetic material and the intersection defines the *operating point* of the system. The load line can be produced by applying the Ampère circuital law (Eq. (4.15)) to an Amperian loop which follows a field line around the magnetic circuit,

$$\oint_C \mathbf{H}.\mathbf{dr} = H_m(2\pi R - l_g) + H_g l_g = 0 \tag{4.48}$$

We should note that $I = 0$ as there is no current-carrying coil. We are also using the fact that the magnitude of the magnetic field inside the magnetic material H_m is uniform and parallel to $\mathbf{dr}$ at all points around the loop, and the magnetic field in the air gap H_g is also uniform, but will be different from that in the magnetic material. An equation that relates H_m to the magnetic flux density in the magnetic material B_m is required, but Eq. (4.48) relates H_m and H_g. However, we know that

$$B_g = \mu_0 H_g \tag{4.49}$$

in the air gap, and that magnetic flux is conserved between materials. If we assume that the magnetic flux lines do not spread out in the air gap because l_g is small compared with r, then this conservation of flux tells us that

$$B_g = B_m \tag{4.50}$$

Substituting Eqs. (4.49) and (4.50) into Eq. (4.48) gives

$$H_m(2\pi R - l_g) + \frac{B_m l_g}{\mu_0} = 0$$

$$B_m = \frac{\mu_0}{l_g}(l_g - 2\pi R)H_m \tag{4.51}$$

This linear equation can be plotted on top of the nonlinear curve relating B_m and H_m for the magnetic material to give the operating point, and an example of this is shown in Figure 4.10. It should be noted that as $l_g \ll 2\pi R$, Eq. (4.51) will produce a line of negative gradient which passes through the origin. Therefore, the operating point will be in the top left-hand quadrant of the magnetization curve, as discussed for hard ferromagnetic materials in Section 4.5.

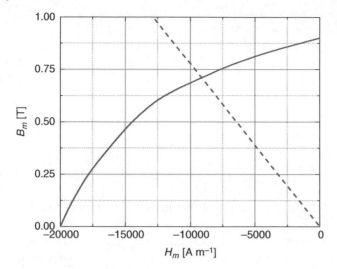

Figure 4.10 The solid line shows the relation between the magnetic field H and magnetic flux density B for 35% cobalt steel. Superposed is the load line of Eq. (4.51) for a toroid with a radius $R = 5$ cm and an air gap of length $l_g = 5$ mm. The intersection of the two lines indicates the actual operating point of the system. This shows that the magnetic flux density is 0.71 T.

In general, magnetic circuit problems involving combinations of nonlinear and linear magnetic materials can be solved by combining three pieces of information: (i) the magnetization curve for the nonlinear material, (ii) the Ampère circuital law to generate an equation that relates H in the different materials, and (iii) the conservation of flux.

The magnetic circuit can be coupled with the concept of *virtual work* to allow the force generated between magnetized objects to be calculated. Perhaps one of the earliest 'experiments' that we all perform as children is to play with magnets, observing how they snap together and then require a significant effort to start to pull apart, but once separated they move apart easily.

By way of an example, let us consider a simple horseshoe magnet made from Alnico (a common hard ferromagnetic iron alloy with aluminium, nickel, cobalt and small quantities of copper) which has a simple cast steel keeper across its ends, as shown in Figure 4.11, where the cross-sectional area A of the horseshoe and the keeper is the same so that the magnetic flux density is also the same all round the magnetic circuit. From the relation between the magnetic field H and the magnetic flux density B for the Alnico shown in Figure 4.12, it is clear that the magnetic flux density in the circuit will not exceed 1.2 T. From Figure 4.8, we can see that there is a fairly linear relationship between B and H in the cast steel for flux densities up to $\sim$1.2 T, allowing a relative permeability to be defined of $\mu_{rk} = 2000$. By considering an Amperian loop which follows a field line around the magnetic circuit, application of the Ampère circuital law (Eq. (4.15)) gives

$$\oint_C \mathbf{H.dr} = H_m l_m + H_k l_k = 0 \tag{4.52}$$

where H_m and H_k are the magnetic fields in the horseshoe Alnico magnet and the cast steel keeper respectively, and l_m and l_k are the magnetic path lengths in the two materials.

Figure 4.11 The magnetic flux lines passing round a horseshoe magnet attached to a cast steel keeper. The magnetic path lengths in the two materials are l_m and l_k respectively, and both have the same cross-sectional area A.

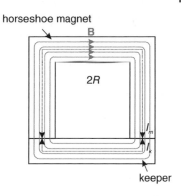

Figure 4.12 A typical graph showing the approximate relation between the magnetic field H and the magnetic flux density B for an Alnico permanent magnetic material, which is a hard ferromagnetic material.

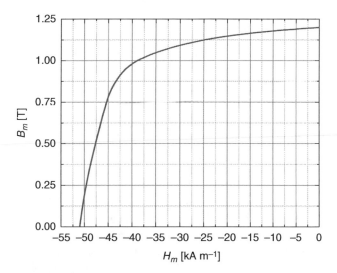

Knowing that the magnetic flux densities in the two materials are the same ($B_m = B_k$), we can rewrite Eq. (4.52) to directly relate the magnetic field and magnetic flux density in the Alnico horseshoe as

$$B_m = \frac{-\mu_0 \mu_{rk} l_m}{l_k} H_m \tag{4.53}$$

Substitution of any sensible number for magnetic path lengths into this equation shows that the operating point is at a very low magnetic field ($\ll 1000 \, \text{A m}^{-1}$), and so the flux density around the circuit will be approximately the remnance flux density of 1.2 T.

Having worked out the magnetic flux density B in the circuit, let us now consider why the keeper is being held in place by the magnet by imagining what would happen if the two were separated by a small distance δx. In order to maintain the magnetic circuit, the magnetic flux lines will have to cross the small gap, so that the magnetic field exists in a small volume of air. We saw in Section 4.6 that it costs energy for a field to occupy a volume of space. In this case, the volume of air in which the field will exist is $2A\delta x$, and so application of Eq. (4.41) allows us to calculate the change in energy δU as a result of opening this gap to be

$$\delta U = \int_v \frac{B^2}{2\mu_0\mu_r} dv = \frac{B^2 A \delta x}{\mu_0} \tag{4.54}$$

remembering that the relative permeability of air is 1. Work has to be done to create this small air gap, and basic mechanics tells us that this requires a force F to be applied over the distance δx such that

$$\delta U = F\delta x \tag{4.55}$$

Therefore, equating Eqs. (4.54) and (4.55) gives an expression for the force acting to hold the magnetic keeper in place of

$$F = \frac{B^2 A}{\mu_0} \tag{4.56}$$

which we would need to apply in order to separate the keeper from the horseshoe magnet. This method of calculation is called *virtual work*, as in reality the keeper does not leave the horseshoe, but by imagining such a virtual movement, we can evaluate the virtual work that would need to be done and hence the force that is acting.

It is notable that the actual separation between the keeper and the horseshoe magnet does not appear in Eq. (4.56). We could repeat the virtual work calculation for a scenario where the two are already separated by a distance x and we are increasing this by a further δx. This is because we are assuming that the flux lines pass straight across the air gap to minimize the energy cost of filling this additional volume of space. This is valid because Eq. (4.40) shows that it represents a much lower energy per unit volume for the flux lines to stay in the cast steel which has a high μ_r. Therefore, the force between the keeper and horseshoe magnet will indeed remain fairly constant until a critical separation is reached when it will no longer be the lowest-energy state for the flux lines to pass through the keeper. Instead, it will be a lower energy for them to occupy a smaller volume of air and pass directly between the two ends of the horseshoe magnet. Suddenly, the force attracting the keeper will be lost. This is why magnets snap together suddenly as such a critical point is reached with a sudden increase in force, just as we observed as children. Likewise, it is this phenomenon that makes magnets so useful as clasps, locks and fasteners as well as in children's toys!

References

Curie, P. (1895). Lois expérimentales du magnétisme. Propriétés magnétiques des corps à diverses températures. *Annales de Chimie et de Physique* 5: 289–405.

Meissner, W. and Ochsenfeld, R. (1933). Ein neuer Effekt bei Eintritt der Supraleitfähigkeit. *Naturwissenschaften* 21 (44): 787–788.

5

The Maxwell Equations of Elecromagnetism

5.1 Introduction to the Maxwell Equations

In 2004, the readers of *Physics World* magazine were posed the question, 'Which equations are the greatest?' The Maxwell equations topped the poll. They are a set of four equations which each describe some aspect of the basic physical properties and origins of electric and magnetic fields in a mathematical form which is elegant in its simplicity. Each can be described in two ways – using integrals and using vector calculus – and we shall see that both have their particular strengths when trying to tackle real problems. As we shall see in the second half of this book, we can use mathematics to manipulate these equations in combination with each other to gain insight into how the electromagnetic world works. For example, they directly predict that electromagnetic waves can propagate through dielectric (insulating) media and give the velocity of the wave as well. Without the Maxwell equations, we would not be able to engineer things as diverse as radio transmitters and anti-reflection coatings on reading glasses.

Despite the power and elegance of the Maxwell equations, students often find them hard to appreciate. This is usually because they can appear to be very abstract. Therefore, in this chapter, we shall consider each of the four equations in turn, in both integral and vector calculus (normally called differential) forms. However, in doing so, we shall try to hold on to the underlying physical phenomena that they try to explain while letting mathematics do some work for us at a more abstract level. At the end of the chapter we shall bring the four equations together so that we can then apply them in the second half of the book, along with the understanding that we have gained in the preceding chapters.

5.2 The Gauss Law of Electric Fields

We have already met the first of the Maxwell equations in Chapters 1 and 2. In Section 1.4, we saw that we could express the fact that charge produces an electric field $\mathbf{E}$ in free space as

$$\oint_S \mathbf{E} \cdot \mathbf{dA} = \frac{q}{\varepsilon_0} \tag{1.12}$$

Although this appears to be a rather complicated integral, in fact the left-hand side of the equation is simply counting the total number of lines of electric flux that pass through a

Electromagnetism for Engineers, First Edition. Andrew J. Flewitt.
© 2023 John Wiley & Sons Ltd. Published 2023 by John Wiley & Sons Ltd.
Companion website: www.wiley.com/go/flewitt/electromagnetism

closed surface over which we are integrating, with lines leaving the closed surface counting positively and those entering counting negatively. As lines of electric flux begin on positive charges and end on negative charges, the only way that the integral can yield a non-zero result is if a flux line has either begun or ended inside the surface, implying that there must be charge enclosed. Then, in Chapter 2, we saw that when an electric field impinges on a dielectric medium, it can induce a polarization which modulates the electric field inside the medium, and we defined the relative permittivity ε_r as a means of quantifying this (Section 2.2). This led to the concept of the electric flux density **D**, which is related to **E** by

$$\mathbf{D} = \varepsilon_0 \varepsilon_r \mathbf{E} \tag{2.15}$$

and consequently, this produced a more general form of Eq. (1.12)

$$\oint_S \mathbf{D.dA} = q_f \tag{2.14}$$

which is valid in all media, where q_f is the free charge enclosed by the closed surface (allowing us to ignore any polarization charges which would make calculations significantly more complex). This is the Gauss law of electric fields in integral form, and the first of the four Maxwell equations.

One of the fundamental operators in vector calculus is the *gradient operator*, which is given the symbol ∇. Commonly called simply *grad*, it can act on any scalar quantity, for example some arbitrary quantity ψ, and in Cartesian coordinates is expressed as

$$\nabla \psi = \left(\frac{\partial \psi}{\partial x}, \frac{\partial \psi}{\partial y}, \frac{\partial \psi}{\partial z} \right) \tag{5.1}$$

The result is a vector quantity where the magnitude of the vector at any point is the magnitude of the gradient in ψ at that point, and the direction is down the line of maximum gradient. It is for this reason that in Section 1.3 we were able to relate electric potential V and electric field **E** in three-dimensional space by

$$\mathbf{E} = -\nabla V \tag{1.7}$$

A second key vector calculus operator which follows from the grad operator is the *divergence operator*, commonly called *div*. This is the dot product of the grad operator with a vector field. For some arbitrary vector field **Y**, it is expressed in Cartesian coordinates as

$$\nabla \mathbf{.Y} = \frac{\partial Y_x}{\partial x} + \frac{\partial Y_y}{\partial y} + \frac{\partial Y_z}{\partial z} \tag{5.2}$$

where Y_x, Y_y and Y_z are the components of **Y** along each Cartesian direction. The physical interpretation of divergence is that it is the net flux of the vector field out of a small volume element at a particular point in space. Therefore, we could work out the total flux of a field leaving an arbitrary volume V of space, by integrating the divergence of the field over that volume, $\int_V (\nabla \mathbf{.Y}) dV$. However, we have already understood that we can calculate the same total flux by using $\oint_S \mathbf{Y.dA}$, where the integral is over the closed surface enclosing the volume. This is the principal behind the *Gauss theorem for divergence*, a full derivation of which can be found in many undergraduate mathematics text books, such as Stroud (1996). This simply equates these two quantities, so

$$\int_V (\nabla \mathbf{.Y}) dV = \oint_S \mathbf{Y.dA} \tag{5.3}$$

Hence, the integral form of the Gauss law of electric fields (Eq. (2.14)) can be rewritten using the Gauss theorem for divergence as

$$\int_V (\nabla.\mathbf{D})dV = q_f \tag{5.4}$$

If there is some arbitrary distribution of free charge inside the volume, then we could define a free *charge density* ρ which is a function of spatial coordinates and has units of $C\,m^{-3}$. The total free charge would then be the integral of the charge density over the volume,

$$q_f = \int_V \rho dV \tag{5.5}$$

Equating Eqs. (5.4) and (5.5) then gives

$$\int_V (\nabla.\mathbf{D})dV = \int_V \rho dV \tag{5.6}$$

and so differentiating both sides of this expression gives the Gauss law of electric fields in differential form,

$$\nabla.\mathbf{D} = \rho \tag{5.7}$$

This should make intuitive sense if we remember that the Gauss law of electric fields is simply a mathematical way of expressing the fact that charge produces an electric field (lines of electric flux begin on positive charges and end on negative charges). Therefore, if there is a non-zero net flux leaving a small volume (given by $\nabla.\mathbf{D}$) then there must be a non-zero local charge density ρ.

5.3   The Gauss Law of Magnetic Fields

In Section 3.6, we saw that there is also a Gauss law of magnetic fields which is the second of the four Maxwell equations, and is expressed in integral form as

$$\oint_S \mathbf{B}.\mathbf{dA} = 0 \tag{3.22}$$

This is a mathematic expression of the phenomenon that there are only magnetic dipoles and no monopoles. Therefore, lines of magnetic flux form closed loops – there are no monopoles upon which lines of flux can either begin or end.

Equation (3.22) has a very similar form to the Gauss law of electric fields in integral form (Eq. (2.14)), and so we can again rewrite this equation using the Gauss theorem for divergence (Eq. (5.3)) as

$$\int_V (\nabla.\mathbf{B})dV = 0 \tag{5.8}$$

Differentiating both sides of this equation with respect to volume gives the Gauss law of magnetic fields in differential form,

$$\nabla.\mathbf{B} = 0 \tag{5.9}$$

This says mathematically that there can never be a net magnetic flux leaving a volume of space, as the magnetic flux lines have no beginning or end, and so any line that enters a volume of space must also leave it.

5.4   The Faraday Law of Magnetic Fields

The third of the four Maxwell equations is based on the Faraday law of electromagnetic induction, which we discussed previously in Section 3.7, and is given by

$$V = -\frac{d\Phi}{dt} \tag{3.26}$$

It describes the phenomenon that a changing magnetic flux through a coil of wire induces a potential difference across the end of the wire that acts to oppose the changing flux.

Let us imagine the scenario shown in Figure 5.1 where a wire carrying a time-varying current I is creating a time-varying magnetic flux density $\mathbf{B}$ which surrounds the wire. This will induce a potential difference V in a second coil of wire which follows the closed loop path indicated by C in Figure 5.1. We know that there must be an electric field present if there is a potential difference from Eq. (1.8). If we say that the closed loop is made up of many small line vectors $\mathbf{dr}$ (as described in Section 3.3), then we can apply Eq. (1.8) as the integral of the electric field around the closed loop must be the potential difference, so

$$\oint_C \mathbf{E}.\mathbf{dr} = V \tag{5.10}$$

We can therefore equate Eq. (5.10) with Eq. (3.26) to give

$$\oint_C \mathbf{E}.\mathbf{dr} = -\frac{d\Phi}{dt} \tag{5.11}$$

However, we also know that the total magnetic flux through the loop at any moment in time Φ is just the integral of the magnetic flux density over the surface S bounded by the closed loop,

$$\Phi = \int_S \mathbf{B}.\mathbf{dA} \tag{5.12}$$

Hence, substitution into Eq. (5.11) gives the *Faraday law of magnetic fields* in integral form

$$\oint_C \mathbf{E}.\mathbf{dr} = -\int_S \frac{d\mathbf{B}}{dt}.\mathbf{dA} \tag{5.13}$$

which is the third Maxwell equation.

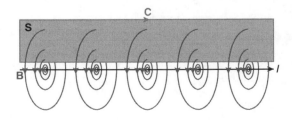

Figure 5.1   A line of wire carrying a current *I* produces a magnetic flux density *B*. A second coil of wire follows the path C and bounds a surface *S* through which the flux passes.

We have already considered the origin of the Faraday law of electromagnetic induction in Section 3.7, and this new expression in Eq. (5.13) which is derived from it is consistent with this earlier understanding. The right-hand side of Eq. (5.13) will only be non-zero if there is a changing magnetic flux passing through the surface bounded by the closed line integral of the left-hand side of Eq. (5.13), meaning that there must be a net electric field circulating round this loop. Therefore, a changing magnetic flux density produces a circulating electric field. This is a key concept. The Gauss law of electric fields (see Section 5.2) makes it clear that lines of electric flux can only begin or end on charges. As there is no charge associated with a time-varying magnetic field, the flux lines of electric field that result cannot have a beginning or end – they must be closed loops, which is why the integral of **E.dr** round the closed loop is non-zero.

The fundamental property that a changing magnetic field produces a circulating electric field, which is expressed mathematically by the Faraday law of magnetic fields, is even more clearly seen in the differential form of this equation. This relies of a third key operator of vector calculus which, like the divergence operator, also follows from the grad operator. The *curl* operator is the cross product of the grad operator with a vector field. For some arbitrary vector field **Y**, it is expressed in Cartesian coordinates as

$$\nabla \times \mathbf{Y} = \begin{vmatrix} \mathbf{i} & \mathbf{j} & \mathbf{k} \\ \partial/\partial x & \partial/\partial y & \partial/\partial z \\ Y_x & Y_y & Y_z \end{vmatrix} \qquad (5.14)$$

where **i**, **j** and **k** are the unit vectors along each of the Cartesian directions and Y_x, Y_y and Y_z are the components of **Y** along each Cartesian direction.

The physical interpretation of curl is that it represents the circulation of a vector field in a small volume element. A way of visualizing this is to imagine a small paddle wheel (such as is found on an old-fashioned paddle steam ship) in a fluid flow represented by a vector field. If the paddle wheel is immersed inside a uniform fluid flow, then it will not rotate as the torques acting on opposite sides of the wheel will always balance, and such a vector field has zero curl. However, if the flow is faster on one side of the wheel compared to the other, such as would be found in a vortex, then the wheel will spin, and this shows that the vector field has a non-zero curl.

An example of a field with a zero curl is the electric field surrounding a point charge, as shown in Figure 1.1. Such a field is clearly divergent but has no circulating element. However, we have seen that the electric field produced by a changing magnetic field is circulatory in nature.

The *Stokes theorem* relates the line integral of an arbitrary vector field **Y** around a closed loop to the surface integral of the vector field through the same loop as

$$\oint_C \mathbf{Y}.\mathbf{dr} = \int_S (\nabla \times \mathbf{Y}).\mathbf{dA} \qquad (5.15)$$

A full derivation of the Stokes theorem may be found in most undergraduate mathematics text books, such as Stroud (1996). The interpretation of the Stokes theorem is that integrating $(\nabla \times \mathbf{Y}).\mathbf{dA}$ over a surface allows us to determine the net circulation of the vector field **Y** over the whole surface. If this is non-zero, then the line integral of the same vector field around the edge of the same surface must also be non-zero.

Application of the Stokes theorem to the integral form of the Faraday law of magnetic fields (Eq. (5.13)) allows the line integral of the electric field to be transformed into a surface integral of the curl of the electric field:

$$\int_S (\nabla \times \mathbf{E}).\mathbf{dA} = -\int_S \frac{d\mathbf{B}}{dt}.\mathbf{dA} \tag{5.16}$$

Differentiating both sides of this equation gives the Faraday law of magnetic fields in differential form:

$$\nabla \times \mathbf{E} = -\frac{\partial \mathbf{B}}{\partial t} \tag{5.17}$$

Therefore, this form of the third Maxwell equation does indeed make it very clear that a locally time-varying magnetic field produces a circulating electric field.

The fact that an electric field can be produced, without the need for free charge, by a time-varying magnetic field lies at the heart of why it is possible for electromagnetic waves to propagate through free space, where clearly there is no charge. We shall return to this in Chapter 7 when we will see that it is possible to use this equation in the derivation of the proof of the existence of electromagnetic waves.

5.5 The Ampère–Maxwell Law

In Section 4.2, we saw the physical phenomenon that a moving charge, such as a current flow, produces a magnetic field. This circulates around the moving charge and can be expressed mathematically using the Ampère circuital law,

$$\oint_C \mathbf{H}.\mathbf{dr} = I \tag{4.15}$$

The integral is around a closed Amperian loop of the dot product of the magnetic field $\mathbf{H}$ and a small vector element of the loop $\mathbf{dr}$, and I is the total current passing through the loop.

In the preceding sections of this chapter, we have seen that it is possible to convert the integral forms of the Maxwell equations into differential forms using vector calculus. We can attempt to do the same thing in this case, again using the Stokes theorem (Eq. (5.15)). This allows the left-hand side of Eq. (4.15) to be rewritten as

$$\oint_C \mathbf{H}.\mathbf{dr} = \int_S (\nabla \times \mathbf{H}).\mathbf{dA} \tag{5.18}$$

where the right-hand side of the equation is the integral over a surface bounded by the closed loop. However, the total current through the loop I on the right-hand side of Eq. (4.15) is just the integral of the current density $\mathbf{J}$ through the surface bounded by the loop:

$$I = \int_S \mathbf{J}.\mathbf{dA} \tag{5.19}$$

We should note here that the current density is a vector quantity as it has both a magnitude and direction at any point. Therefore $\mathbf{J}.\mathbf{dA}$ is the amount of current passing perpendicularly though a small element of surface $\mathbf{dA}$ (see Section 1.4). Substituting Eqs. (5.18) and (5.19) into Eq. (4.15) allows the Ampère law to be rewritten as

$$\int_S (\nabla \times \mathbf{H}).\mathbf{dA} = \int_S \mathbf{J}.\mathbf{dA} \tag{5.20}$$

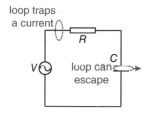

loop traps
a current

R

V loop can
escape

C

Figure 5.2   The need for a displacement current term in the Ampère law is highlighted by this circuit consisting of an a.c. voltage source, a resistor R and a capacitor C. Placing the Amperian loop around most of the circuit will result in a current flow through the loop, and a calculation that a magnetic field exists around the circuit. However, if the loop is placed between the plates of the capacitor, then there is no current passing through the loop, which could be removed from the circuit without changing the current through it. There is clearly an inconsistency.

and differentiating both sides of this equation gives the differential form of the Ampère law,

$$\nabla \times \mathbf{H} = \mathbf{J} \tag{5.21}$$

Maxwell, however, saw an inconsistency with the Ampère law, which is exemplified by the simple electric circuit shown in Figure 5.2, which consists of an a.c. voltage source, a resistor and a capacitor connected in series. A current flows in the circuit to charge and discharge the capacitor. Therefore, if we were to imagine placing a closed Amperian loop around a wire in the circuit and apply Eq. (4.15), we would have a current passing through the loop, and therefore there would be a magnetic field around the circuit. However, if we choose to place the loop in the gap between the plates of the capacitor we would get a different result. No current actually flows through the dielectric between the capacitor's plates, and so the right-hand side of Eq. (4.15) would be zero. However, if the gap is thin, then there is clearly a magnetic field circulating around the circuit in this region, and the left-hand side of Eq. (4.15) would be non-zero. An extra term needs to be added to the Ampère law, called the *displacement current*.

In order to understand how Eq. (5.21) needs to be changed to account for this displacement current, we start by creating a *continuity equation for charge conservation* based on the fact that charge cannot be created or destroyed. Let us consider a small volume of space V which contains a total charge Q. We could calculate this total charge by integrating the charge density ρ over the volume:

$$Q = \int_V \rho \, dV \tag{5.22}$$

If the total charge inside the small volume is changing, then differentiating Eq. (5.22) with respect to time gives

$$\frac{dQ}{dt} = \int_V \frac{\partial \rho}{\partial t} \, dV \tag{5.23}$$

There must be a current flow through the closed surface S bounding the volume, as charge cannot be created or destroyed. We can calculate this current by integrating the current density over this surface,

$$\frac{dQ}{dt} = -\oint_S \mathbf{J.dA} \tag{5.24}$$

and equating this to Eq. (5.23) gives

$$-\oint_S \mathbf{J}.\mathbf{dA} = \int_V \frac{\partial \rho}{\partial t} dV \tag{5.25}$$

We can apply the divergence theorem (Eq. (5.3)) to the left-hand side of this equation,

$$-\int_V (\nabla.\mathbf{J}) dV = \int_V \frac{\partial \rho}{\partial t} dV \tag{5.26}$$

Finally, we can differentiate both sides with respect to volume to give

$$\nabla.\mathbf{J} = -\frac{\partial \rho}{\partial t} \tag{5.27}$$

Remembering that the physical interpretation of divergence is that it is the net flux of a quantity out of a small volume element, then in this case a non-zero divergent current must mean that there is a similarly non-zero rate of change of charge in the volume. Equation (5.27), therefore, is the continuity equation for charge conservation. This can be compared with the continuity equation for mass conservation in a fluid flow, which is

$$\nabla.(\rho_m \mathbf{v}) = -\frac{\partial \rho_m}{\partial t} \tag{5.28}$$

In this case, ρ_m is the mass density and $\mathbf{v}$ is the velocity of the fluid flow. Therefore, $\rho_m \mathbf{v}$ is really the mass flux (in a similar way that current density is a charge flux), and so a divergent mass flux must mean a change in total mass, and therefore mass density, within the volume.

Let us now return to the issue of the displacement current that is missing from Eq. (5.21). From the differential form of the Gauss law of electric fields (Eq. (5.7)), we know that $\nabla.\mathbf{D} = \rho$, so substitution into the continuity equation for charge conservation (Eq. (5.27)) gives

$$\nabla.\mathbf{J} = -\nabla.\left(\frac{\partial \mathbf{D}}{\partial t}\right) \tag{5.29}$$

However, if Eq. (5.21) is correct then, by taking the divergence of both sides,

$$\nabla.\mathbf{J} = \nabla.(\nabla \times \mathbf{H}) = 0 \tag{5.30}$$

as there is a vector calculus identity that for any arbitrary vector field $\mathbf{Y}$,

$$\nabla.(\nabla \times \mathbf{Y}) = 0 \tag{5.31}$$

Equation (5.30) (and by implication Eq. (5.21)) is clearly inconsistent with Eq. (5.29), which has come directly from the continuity equation for charge conservation. We therefore call $\partial \mathbf{D}/\partial t$ the displacement current (on account of it being dimensionally the same as current) and we add this to the current density in Eq. (5.21) to give the differential form of the *Ampère–Maxwell law*,

$$\nabla \times \mathbf{H} = \mathbf{J} + \frac{\partial \mathbf{D}}{\partial t} \tag{5.32}$$

which is the last of the four Maxwell equations. Taking the divergence of both sides of this equation gives Eq. (5.29), and so there is now consistency with the continuity equation for charge conservation.

It is possible to produce an integral version of the Ampère–Maxwell law by taking the integral of Eq. (5.32) over a surface S that is bounded by a closed loop C:

$$\int_S (\nabla \times \mathbf{H}).\mathbf{dA} = \int_S \mathbf{J}.\mathbf{dA} + \int_S \frac{\partial \mathbf{D}}{\partial t}.\mathbf{dA} \tag{5.33}$$

Substituting for Eqs. (5.18) and (5.19) then gives the integral form as

$$\oint_C \mathbf{H}.\mathbf{dr} = I + \int_S \frac{\partial \mathbf{D}}{\partial t}.\mathbf{dA} \tag{5.34}$$

The physical interpretation of the Ampère–Maxwell equation is that a circulating magnetic field can be produced in either of two ways: by a moving charge (a current density) or by a time-varying electric flux density. It is this latter mechanism that is new, but the symmetry with the Faraday law of magnetic fields in differential form (Eq. (5.17)) should be noted, as this states that a circulating electric field can be produced by a time-varying magnetic field. As we shall see in Chapter 7, it is the combination of these two mechanisms that effectively allows electromagnetic waves to propagate through free space.

5.6   Work Done to Produce an Electromagnetic Field

In considering the fundamental nature of electric and magnetic fields in the first part of this book, we have noted several times that a field occupying a volume of space v has an energy given by

$$U = \int_v \int_0^E \mathbf{D}.d\mathbf{E}\, dv \tag{2.27}$$

for electric fields and

$$U = \int_v \int_0^H \mathbf{B}.d\mathbf{H}\, dv \tag{4.40}$$

for magnetic fields. However, we have not derived where these two important expressions come from. In order to do this, it is better to solve the problem of how much energy is required to produce such fields in the first place.

Let us consider the case for electric fields first. We defined potential difference V in Section 1.3 to be the change in energy per unit charge between two points in space. Therefore, we can determine the small change in energy δU required to change the charge density by a small amount in a volume of space v if this volume is at some potential V relative to the potential an infinite distance away. In effect, we are saying that we have brought the charge required to change the charge density from an infinite distance away to this volume. From the definition of potential difference, this small change in energy is

$$\delta U = \int_v V \delta\rho\, dv \tag{5.35}$$

By including V inside the integration, we are allowing for the fact that the potential may not be a constant everywhere in the volume. We can relate the small change in charge density to an associated small change in electric flux density $\delta \mathbf{D}$ using the Gauss law of electric fields (Eq. (5.7)) as

$$\delta U = \int_v V(\nabla.\delta\mathbf{D})dv \tag{5.36}$$

We now need to use the vector calculus identity that, for some arbitrary vector field $\mathbf{Y}$ and a scalar quantity φ that is a function of position,

$$\nabla.(\varphi\mathbf{Y}) = (\mathbf{Y}.\nabla)\varphi + \varphi(\nabla.\mathbf{Y}) \tag{5.37}$$

Therefore, we can rewrite Eq. (5.36) as

$$\delta U = \int_v [\nabla.(V\delta \mathbf{D}) - (\delta \mathbf{D}.\nabla)V]dv \qquad (5.38)$$

We can now use the Gauss theorem for divergence to convert the first term in the volume integral into a surface integral:

$$\delta U = \oint_S V\delta \mathbf{D}.\mathbf{dA} - \int_v (\delta \mathbf{D}.\nabla)Vdv \qquad (5.39)$$

As we can choose what volume we wish to consider, we can pick a situation where the volume extends a long distance around the small charge that we have added, and therefore the surface integral term will tend to zero, leaving

$$\delta U = -\int_v \delta \mathbf{D}.\nabla Vdv \qquad (5.40)$$

However, we also know that the electric field and potential are related by

$$\mathbf{E} = -\nabla V \qquad (1.7)$$

and so substitution into Eq. (5.40) gives

$$\delta U = \int_v \delta \mathbf{D}.\mathbf{E}\, dv \qquad (5.41)$$

We can now use the relationship $\mathbf{D} = \varepsilon_0 \varepsilon_r \mathbf{E}$ (Eq. (2.15)) to substitute for both $\delta \mathbf{D}$ and $\mathbf{E}$ in Eq. (5.41) to effectively swap their roles so that

$$\delta U = \int_v \mathbf{D}.\delta \mathbf{E}\, dv \qquad (5.42)$$

Finally, we can integrate the electric field from zero to find the total energy in the field:

$$U = \int_v \int_0^E \mathbf{D}.d\mathbf{E}\, dv \qquad (2.27)$$

In the case of magnetic fields, we need to consider the current density that is producing the field. Applying a current density $\mathbf{J}$ in a short time δt will need work δU to be done against any electric field $\mathbf{E}$. This work is given by

$$\delta U = -\delta t \int_v \mathbf{E}.\mathbf{J}dv \qquad (5.43)$$

The origin of this equation is the familiar expression for power dissipated by a current flowing across a potential difference where $P = VI$. If the current is flowing through an area A and over a distance l then we can relate the current to the current density using $I = JA$ and the voltage to the electric field by $V = El$. However, real power is only dissipated if the current is flowing directly against the electric field, and therefore we can use the vector forms of the current density and electric field and the fact that the dot product is the component of one vector in the direction of the other to give $P = -Al\mathbf{E}.\mathbf{J}$. As Al is the volume in which the power is being dissipated, $-\mathbf{E}.\mathbf{J}$ is just the power dissipated per unit volume, which leads directly to Eq. (5.43).

If we assume that the magnetic flux density is not changing, then from the Ampère–Maxwell law (Eq. (5.32)) we can express the current density in terms of the change in magnetic field as

$$\delta U = -\delta t \int_v \mathbf{E}. \, (\nabla \times \mathbf{H})dv \tag{5.44}$$

We can now use the vector calculus identity for two arbitrary fields $\mathbf{Y}_1$ and $\mathbf{Y}_2$ given by

$$\nabla.(\mathbf{Y}_1 \times \mathbf{Y}_2) = \mathbf{Y}_2.(\nabla \times \mathbf{Y}_1) - \mathbf{Y}_1.(\nabla \times \mathbf{Y}_2) \tag{5.45}$$

and so Eq. (5.44) becomes

$$\delta U = \delta t \int_v [\nabla.(\mathbf{E} \times \mathbf{H}) - \mathbf{H}.(\nabla \times \mathbf{E})]dv \tag{5.46}$$

Just as with the electric field case, we can use the Gauss theorem for divergence to convert the first of the terms in Eq. (5.46) into a surface integral which tends to zero for large surfaces a long distance from the current density, and so this reduces to

$$\delta U = -\delta t \int_v \mathbf{H}.(\nabla \times \mathbf{E})dv \tag{5.47}$$

We can now use the Faraday law of magnetic fields (Eq. (5.17)) to convert the electric field term into a magnetic flux density

$$\delta U = \delta t \int_v \mathbf{H}. \left(\frac{d\mathbf{B}}{dt} \right) dv \tag{5.48}$$

Next, we can cancel the time differential terms and use the relationship $\mathbf{B} = \mu_0 \mu_r \mathbf{H}$ (Eq. (2.15)) to substitute for both $\mathbf{H}$ and $\mathbf{B}$ to give

$$\delta U = \int_v \mathbf{B}.d\mathbf{H} \, dv \tag{5.49}$$

Finally, we can integrate the magnetic field from zero to find the total energy in the field:

$$U = \int_v \int_0^H \mathbf{B}.d\mathbf{H} \, dv \tag{4.40}$$

These two derivations show that energy is stored in electromagnetic fields and they allow us to calculate how much energy is stored. We shall return to this idea when we consider the power in electromagnetic waves in Section 7.7.

5.7 Summary of the Maxwell Equations

Over the course of this chapter, we have looked in detail at each of the four Maxwell equations in differential form, and how they may be determined from their integral equivalents. These are now summarized along with their physical interpretation in Table 5.1.

Table 5.1 Summary of the Maxwell equations.

Law	Integral form	Differential form	Interpretation
Gauss law of electric fields	$\oint_S \mathbf{D}.\mathbf{dA} = q_f$	$\nabla.\mathbf{D} = \rho$	Charge produces an electric field with lines of flux beginning on positive and ending on negative charge
Gauss law of magnetic fields	$\oint_S \mathbf{B}.\mathbf{dA} = 0$	$\nabla.\mathbf{B} = 0$	There are only magnetic dipoles so lines of magnetic flux form closed loops
Faraday law of magnetic fields	$\oint_C \mathbf{E}.\mathbf{dr} = -\int_S \frac{d\mathbf{B}}{dt}.\mathbf{dA}$	$\nabla \times \mathbf{E} = -\frac{\partial \mathbf{B}}{\partial t}$	A time-varying magnetic flux density induces a circulating electric field
Ampère-Maxwell law	$\oint_C \mathbf{H}.\mathbf{dr} = I + \int_S \frac{\partial \mathbf{D}}{\partial t}.\mathbf{dA}$	$\nabla \times \mathbf{H} = \mathbf{J} + \frac{\partial \mathbf{D}}{\partial t}$	A moving charge or a time-varying electric flux density produce a circulating magnetic field

Reference

Stroud, K.A. (1996). *Further Engineering Mathematics: Programmes and Problems*. Basingstoke: Macmillan.

Part II

Applications of Electromagnetism

6

Transmission Lines

6.1 Introduction

In considering the physical meaning of both electric and magnetic fields in Part 1 of this book, we saw that energy is stored whenever a field occupies a volume of space (see Section 5.6). Therefore, if we can make an electric or magnetic field move through space, then it follows that energy must also be moving as well. We also know that we can create an electric field by spatially separating positive and negative charges, for example by applying a potential difference between two spatially separated conductors, and we can create a magnetic field by allowing a current to flow along a conductor. Transmission lines are a consequence of this. They allow energy to be transmitted from one point in space to another by connecting the two points using two or more conductors which then allow a moving electric and magnetic field, usually called an electromagnetic wave, to travel between the two points.

Transmission lines take many forms and can operate over a wide range of length scales, as indicated in Figure 6.1. At the macro scale are electricity power transmission lines which allow energy to be transferred from a generation plant to remote points of use. This is normally achieved using a low-frequency (typically 50 or 60 Hz) sinusoidal voltage applied between either two or three conductors, which results in an alternating current (a.c.) flow on each conductor. Transmission distances can be very long, easily extending hundreds of kilometres. We shall discuss power transmission in more detail in Chapter 12. On a smaller length scale, we have cables that transmit information encoded as an electrical signal over shorter distances, such as the coaxial cable. In this case, there are two conductors – an inner conductor and an outer conductor separated by an insulating dielectric, as shown in Figure 2.7. Either a digital or analogue voltage signal is applied between the two conductors at one end, which is transmitted down the cable to the other end. A typical use of coaxial cables is to connect a television aerial on the roof of a building to a television. In this case, the sinusoidal voltage signal has a frequency in the region of 400–900 MHz, and we measure the length of the transmission line in metres. At an even smaller length scale, we have the microstrip lines on a printed circuit board (PCB) which connect electronic components together in a simple system, such as on the motherboard of a computer. In this case, there is one conductor connecting one pin on a device to a pin on another device and a common ground plane conductor on the reverse side of the PCB. The signal is the voltage on

Electromagnetism for Engineers, First Edition. Andrew J. Flewitt.
© 2023 John Wiley & Sons Ltd. Published 2023 by John Wiley & Sons Ltd.
Companion website: www.wiley.com/go/flewitt/electromagnetism

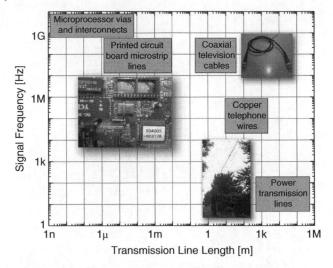

Figure showing signal frequency vs transmission line length with labels: Microprocessor vias and interconnects, Printed circuit board microstrip lines, Coaxial television cables, Copper telephone wires, Power transmission lines. Y-axis: Signal Frequency [Hz] with markings 1, 1k, 1M, 1G. X-axis: Transmission Line Length [m] with markings 1n, 1μ, 1m, 1, 1k, 1M.

Figure 6.1 Examples of transmission lines with indicative line lengths and signal frequencies showing their wide range of uses.

the conducting track with respect to the ground plane, and typically has a frequency up to 800 MHz. The track length is measured in millimetres.

Transmission lines are therefore ubiquitous. In this chapter, we shall look at how transmission lines may be analysed using an electrical equivalent circuit model. This gives us an understanding of exactly why they work and how they can be engineered with the desired properties to make them satisfy any one of these diverse range of applications. We shall also see how to analyse transmission lines when they are connected to real sources of a signal and receiving loads.

6.2   Ideal Transmission Lines and the Telegrapher's Equations

Although at one level it may seem as though there is little similarity between the different types of transmission lines considered in Section 6.1, in fact they all have a simple feature in common: they consist of at least two conductors which run between a source point and a receiving point, and which are separated along their length by a dielectric (insulating) medium, whether that is air, as is the case for the power transmission line, or a plastic as used in a coaxial cable.

Let us imagine that we have some arbitrary transmission line of infinite length, which starts at some point in space which we will define as being $x = 0$, where we have connected a voltage source producing some sinusoidal a.c. signal. We will now try and construct an equivalent electric circuit to model an infinitesimally short length δx of this transmission line at some distance x along its length away from the source, as shown in Figure 6.2.

We know from Section 2.3 that two conductors which are separated by a dielectric have a capacitance which acts between them. Therefore, we can define a capacitance per unit length of the transmission line, C. For example, let us consider a simple microstrip line on a PCB of width w with an underlying dielectric of thickness d, as shown in Figure 6.3.

Figure 6.2 Equivalent circuit of a short length δx of a transmission line at a point a distance x along the line away from an a.c. voltage source connected to the start of the line at $x = 0$. The arbitrary length of line between the source and the short length being considered is represented by a dashed line. A dashed line is also shown to indicate that the line continues to infinity.

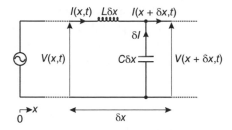

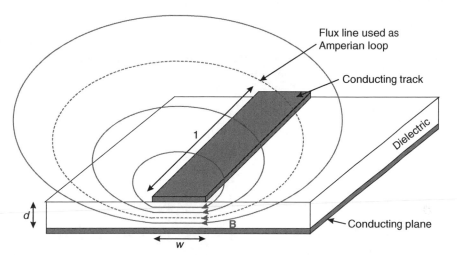

Figure 6.3 Schematic diagram of a microstrip line of width w and unit length on a dielectric of thickness d. The flux lines are shown schematically. One flux line (shown dashed) is chosen as the Amperian loop for application of the Ampère circuital law to calculate the flux density in the region of uniform flux between the conductors.

This will behave like a simple parallel plate capacitor, for which we derived an expression for the capacitance in Section 2.3,

$$C = \frac{\varepsilon_0 \varepsilon_r A}{d}$$

where A is the area between the plates of the capacitor, which for a unit length of the microstrip line will be $A = 1 \times w$, d is the thickness of the dielectric between the conductors and ε_r is its relative permittivity. Therefore, the capacitance per unit length of the microstrip line is

$$C = \frac{\varepsilon_0 \varepsilon_r w}{d} \tag{6.1}$$

Hence, the capacitance of the small length δx of the transmission line is $C\delta x$, and this will look like a capacitor acting between the two conductors, as charge is building up between the conductors, and so it appears in this way in the equivalent circuit of Figure 6.2.

We also know from Section 3.8 that a loop of wire carrying a current will have an inductance, and so as the transmission line is carrying a current, it must have an inductance per unit length L, which we can derive for the example of the microstrip line using the methodology set out in Section 3.8. The magnetic flux lines will circulate around the top conducting line, and because there is a return current path through the bottom conducting plane, they will be confined to be above this plane. As a result the flux lines will be constricted to pass through the small area between the conductors, but will then spread out to minimize energy while still forming closed loops, as shown in Figure 6.3. We will make the assumption that the magnetic field is zero everywhere apart from in the region between the conductors where it has a magnitude of H. Therefore, by applying the Ampère circuital law (Eq. (4.15)) to the Amperian loop shown, noting that as it follows a flux line **H** and **dr** are always parallel, we have

$$\oint_C \mathbf{H}.\mathbf{dr} = Hw = I \tag{6.2}$$

Most dielectrics are ideal, linear magnetic materials, and so $B = \mu_0 \mu_r H$, giving an expression for the magnetic flux density in the region between the conductors

$$B = \frac{\mu_0 \mu_r I}{w} \tag{6.3}$$

If we imagine a unit length of the microstrip line, then the total flux passing through the dielectric in a cross section between the conductors is

$$\Phi = B \times 1 \times d = \frac{\mu_0 \mu_r I d}{w} \tag{6.4}$$

and the inductance per unit length is

$$L = \frac{d\Phi}{dI} = \frac{\mu_0 \mu_r d}{w} \tag{6.5}$$

Hence, the inductance of the small length δx of the transmission line is $L\delta x$, and this will look like an inductor acting along the conductor, as current is flowing along the conductor, and so it appears in this way in the equivalent circuit of Figure 6.2.

We now have our equivalent circuit of the short length δx of our arbitrary transmission line. As the short length has a finite inductance $L\delta x$, at any moment in time t, there will be a different voltage on the left-hand side of the short length $V(x, t)$ compared with the right-hand side $V(x + \delta x, t)$, with the difference

$$\delta V = V(x + \delta x, t) - V(x, t) \tag{6.6}$$

from the application of the Kirchhoff voltage law being the potential difference across the inductor. This can be related to the rate of change of current through the inductor using Eq. (3.39) to give

$$\delta V = -L\delta x \frac{dI}{dt} \tag{6.7}$$

which, in the limit as δV and δx tend to zero, rearranges using basic calculus to

$$\frac{dV}{dx} = -L\frac{dI}{dt} \tag{6.8}$$

Similarly, there will be a current δI flowing through the capacitor, meaning that the current on the left-hand side of the short length $I(x, t)$ will be different compared to that on the right-hand side $I(x + \delta x, t)$, whereby application of the Kirchhoff current law,

$$I(x + \delta x, t) = I(x, t) + \delta I \tag{6.9}$$

The current flowing down the potential difference $V(x, t)$ across the capacitor is therefore $-\delta I$, and this can be related to the rate of change of voltage across the capacitor using Eq. (2.26) to give

$$-\delta I = C\delta x \frac{dV}{dt} \tag{6.10}$$

which, in the limit as δI and δx tend to zero, rearranges using basic calculus to

$$\frac{dI}{dx} = -C\frac{dV}{dt} \tag{6.11}$$

Collectively, Eqs. (6.8) and (6.11) describe the relationship between the current and voltage as a function of position and time on a transmission line and are known as the *telegrapher's equations*. They show very clearly that the inductance and capacitance per unit length of the transmission line quantitatively determine these relationships. The name refers to one of the earliest applications of transmission lines, developed in the nineteenth century, being the transmission of data using electrical signals along wires using protocols like Morse code. Networks of cables were set up which spanned the globe allowing messages, called 'telegrams' or 'cables', to be sent worldwide essentially instantaneously for the first time.

6.3   Waves on Ideal Transmission Lines

The telegrapher's equations (Eqs. (6.8) and (6.11)) tell us how the current and voltage at any point on a transmission line relate to each other as a function of time, but they do not directly tell us how current and voltage evolve with time and position. As we have two equations and two unknowns (voltage and current), we can substitute one into the other to create expressions for voltage or current alone.

Let us differentiate both sides of Eq. (6.8) with respect to x, noting that inductance L is just a constant, so that

$$\frac{d^2V}{dx^2} = -L\frac{d}{dx}\frac{dI}{dt} \tag{6.12}$$

As this is a first-order differentiation, we can change the order on the right-hand side to give

$$\frac{d^2V}{dx^2} = -L\frac{d}{dt}\frac{dI}{dx} \tag{6.13}$$

We can now substitute Eq. (6.11) into Eq. (6.13) to eliminate current,

$$\frac{d^2V}{dx^2} = LC\frac{d^2V}{dt^2} \tag{6.14}$$

Similarly, by differentiating Eq. (6.11) with respect to x and substituting in Eq. (6.8), we can eliminate voltage to give a similar expression to Eq. (6.14) in terms of current only:

$$\frac{d^2I}{dx^2} = LC\frac{d^2I}{dt^2} \tag{6.15}$$

Equations (6.14) and (6.15) both have a similar form: they are one-dimensional *wave equations*.

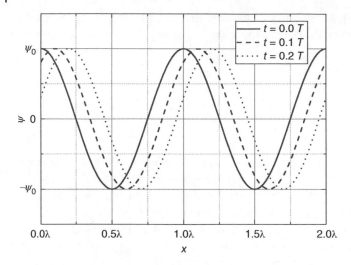

Figure 6.4 A wave of the form $\psi = \psi_0 \cos(\omega t - \beta x)$ plotted over a distance of two wavelengths for three different times within the first time period T, clearly showing that the wave is moving in a positive x-direction.

The one-dimensional wave equation appears in many branches of science and engineering, including longitudinal sound waves in pipes, longitudinal mechanical waves on springs and transverse mechanical waves on strings. In all these cases we find that there are equivalents to voltage and current, as some force (equivalent to the voltage gradient acting on an electron to produce a force) leads to a wave velocity (equivalent to the resulting movement of the electrons to give a current). For example, these quantities are the tension and local velocity at a point respectively for a transverse wave on a string. The wave equation always has the form

$$\frac{d^2\psi}{dx^2} = \frac{1}{c^2}\frac{d^2\psi}{dt^2} \tag{6.16}$$

where ψ is the equivalent of either voltage or current and c is a constant whose physical meaning we shall now deduce.

There are many possible solutions to the one-dimensional wave equation. However, we are normally interested in transmitting sinusoidal a.c. electrical signals down a transmission line, and in any case, we can superpose an infinite series of sinusoidal signals to create any arbitrary wave using a Fourier series (Stroud 1996). Therefore, let us consider

$$\psi = \psi_0 \cos(\omega t - \beta x) \tag{6.17}$$

as a possible solution to Eq. (6.16). This equation represents a wave moving in the positive x-direction with time t, as shown in Figure 6.4, and it has three important constants. ψ_0 is simply the *amplitude* of the wave. ω is the *angular frequency*, and it is related to the *time period* (the time to complete one oscillation) T by

$$\omega = \frac{2\pi}{T} \tag{6.18}$$

Therefore, when $t = T$, it can be seen that $\omega t = 2\pi$ (i.e. one complete temporal cycle of the wave has been completed). Furthermore, as frequency f is the inverse of the time period,

$$\omega = 2\pi f \tag{6.19}$$

Finally, β is the *propagation constant*, and it can be thought of as the spatial equivalent to the angular frequency. The *wavelength* λ is the distance to complete one oscillation (note the similarity with the time period), and it is related to the propagation constant by

$$\beta = \frac{2\pi}{\lambda} \tag{6.20}$$

so that when $x = \lambda$, it can be seen that $\beta x = 2\pi$ (i.e. one complete spatial cycle of the wave has been completed). We should also note that an alternative solution to Eq. (6.16) of the form

$$\psi = \psi_0 \cos(\omega t + \beta x) \tag{6.21}$$

represents a wave moving in the negative x-direction. We can now see that if we substitute Eq. (6.17) into the wave equation (Eq. (6.16)), we get the result

$$\beta^2 \psi = \frac{1}{c^2} \omega^2 \psi \tag{6.22}$$

Hence, the constant c is actually the velocity of the wave, as rearranging Eq. (6.22) gives

$$c = \frac{\omega}{\beta} = \frac{2\pi f}{2\pi / \lambda} = f\lambda \tag{6.23}$$

If we compare the form of the generic one-dimensional wave equation (Eq. (6.16)) with either of the specific wave equations that we have generated for the voltage and current on a transmission line (Eqs. (6.14) and (6.15)), we can see that

$$c = \frac{1}{\sqrt{LC}} \tag{6.24}$$

The velocity of the wave on a transmission line is entirely determined by its inductance and capacitance per unit length. Furthermore, by way of an example, we can substitute the values for L and C that we derived for the microstrip transmission line in Eqs. (6.1) and (6.5) into Eq. (6.24). In this case,

$$LC = \frac{\mu_0 \mu_r d}{w} \times \frac{\varepsilon_0 \varepsilon_r w}{d} = \varepsilon_0 \varepsilon_r \mu_0 \mu_r \tag{6.25}$$

and therefore

$$c = \frac{1}{\sqrt{\varepsilon_0 \varepsilon_r \mu_0 \mu_r}} \tag{6.26}$$

In fact, for all ideal transmission lines, we always obtain this result. ε_r and μ_r are the relative permittivity and relative permeability of the dielectric between the conductors, and so the velocity of the wave on an ideal transmission line is only dependent on the properties of this dielectric, and not on the properties of the conductors. If the dielectric is air or a vacuum, for which $\varepsilon_r = 1$ and $\mu_r = 1$, then $c = \sqrt{8.854 \times 10^{-12} \times 4\pi \times 10^{-7}} = 2.998 \times 10^8 \text{ m s}^{-1}$; that is, the wave travels at the speed of light.

At first sight, this is perhaps a surprising result, as when we considered simple direct current flow in Section 2.6, we saw that the application of a d.c. potential difference between

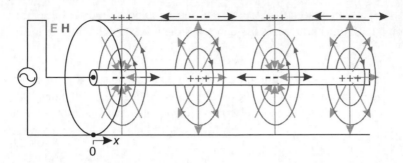

Figure 6.5 A schematic showing the radial electric and circulating magnetic fields at some instant in time on a transmission line. There is local axial movement of charge, as indicated by the arrows pointing along the length of the line.

the ends of a metallic wire sets up an internal axial electric field along its length. However, this is not the case for transmission lines, where the potential difference is applied between the two conductors *at one end* of the transmission line.

To exemplify what is happening, let us take an example of connecting an a.c. voltage source to one end of a coaxial cable, as shown in Figure 6.5. At the start of a sinusoidal voltage cycle, the potential difference applied across the end of the coaxial cable is zero, but will start to increase so that a positive potential difference begins to appear on the inner conductor with respect to the earthed outer conductor. A capacitance exists between the two conductors, and so electrons will be transferred from the inner conductor onto the outer conductor as a current flow through the voltage source. This will have two effects. Firstly, the net positive charge which will exist on the outer surface of the inner conductor and net negative charge on the inner surface of the outer conductor will cause an electric field to exist in the dielectric between the two that will point radially outwards. Secondly, the flow of classical current into the inner conductor will cause a clockwise circulating magnetic field to exist in the dielectric. Both the electric and magnetic fields will be time-varying in magnitude as we are using an a.c. voltage source, and they are clearly perpendicular to each other. As we shall see in Chapter 7, these are the conditions for the propagation of an electromagnetic wave in the direction perpendicular to the electric and magnetic field – the wave will propagate down the length of the transmission line. As it does so, the travelling electromagnetic wave will cause the electrons on the conductors to oscillate locally backwards and forwards along the length of the line, creating a current whose mathematical form is a solution to Eq. (6.15). Critically, as the electromagnetic wave takes some time to travel down the length of the transmission line, there is a phase difference in the electron current with position x. This is analogous to a mechanical longitudinal wave travelling along a slinky spring, as shown in Figure 6.6, where there is a phase difference in local velocity along the length of the spring. The result is that at some points along the spring, the coils are locally compressed, while at other points they are locally extended. In the case of the transmission line, this means that locally the electrons can be more bunched together, producing a local net negative charge, or more spread out, producing a local net positive charge, as shown in Figure 6.5. As lines of electric field begin and end on charges (see Section 1.2), there will be a local electric field. Likewise, the local current flows will produce a circulating magnetic field. The charge in the conductors of the transmission line is effectively guiding the

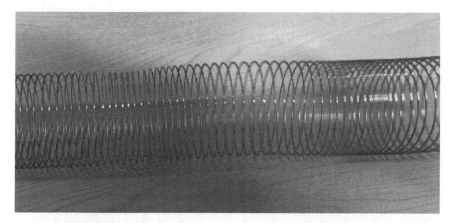

Figure 6.6 A wave on a slinky spring showing regions of local compression and extension.

electromagnetic wave along its path by helping to support the oscillating electric and magnetic fields in the dielectric. Note that in the case of the slinky spring any single point on the spring is only oscillating locally backwards and forwards, but this leads to a wave travelling from end to end of the spring. Likewise in the transmission line any one electron is only oscillating locally backwards and forwards, but this leads to an electromagnetic wave travelling from one end of the line to the other.

You will have experienced the presence of an electromagnetic wave existing around the conductors of a transmission line if you have been listening to the radio in a car when you have driven underneath a power transmission line, as you hear significant interference. Indeed, if you stand under such a line and hold a fluorescent tube vertically upwards to align with the electric field, then the tube will faintly light up, as shown in Figure 6.7. The tube is extracting a small amount of energy from the electromagnetic wave, even though it has no physical connection to the conductors themselves.

You may be wondering why you have never had to consider this wave-like behaviour before when doing basic a.c. circuit analysis. The key quantity to consider is the wavelength, as this determines the distance over which there is a significant phase difference between the local electron current. Figure 6.8 plots the wavelength of an electromagnetic wave in air as a function of frequency. At typical a.c. power frequencies (50 or 60 Hz), the wavelength is well in excess of 1000 km. Therefore, there is negligible phase difference in the current between any two points in the circuit – the current is behaving like an incompressible fluid. However, the wavelength drops to below 10 cm if we go up in frequency to ~3 GHz. This is a typical clock speed in a microprocessor. The clock is a simple square wave signal whose purpose is to provide a timing reference throughout the microprocessor. Given that the physical size of a microprocessor is ~1 cm, attempting to increase the clock speed significantly above several gigahertz would result in a phase difference in the signal around the microprocessor, meaning that it is no longer acting as a good timing reference.

In Eq. (6.17), we saw that a cosine wave is a good general solution to the wave equation. We can use this as the basis for producing a general solution for the voltage on a transmission line, which is a solution to Eq. (6.14). It is possible that the wave could be travelling 'forwards' in the positive x-direction or 'backwards' in the negative x-direction, or indeed

Figure 6.7 Fluorescent tubes illuminated by harvesting energy from the electromagnetic wave around a power transmission line. Source: © James R. Larkin, reproduced with permission from https://www.larkinweb.co.uk/miscellany/fluorescent_tubes_under_power_lines.html.

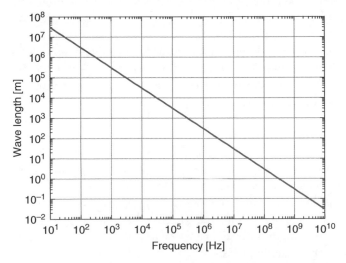

Figure 6.8 The wavelength of an electromagnetic wave in air as a function of frequency calculated using $\lambda = c/f$.

we could have both cases simultaneously. We can express this mathematically as

$$V = V_F \cos(\omega t - \beta x + \phi_F) + V_B \cos(\omega t + \beta x + \phi_B) \tag{6.27}$$

where $V_F \cos(\omega t - \beta x + \phi_F)$ is the forward wave and $V_B \cos(\omega t + \beta x + \phi_B)$ is the backward wave, the direction being determined by the sign of the βx term in each case. V_F and V_B

are then the amplitudes of the forward and backwards waves respectively, while ϕ_F and ϕ_B are the phase of the forward and backward waves respectively at $t = 0$ and $x = 0$. As for a.c. circuit analysis, it is more convenient to express the voltage using complex notation, where from the de Moivre theorem, Eq. (6.27) is the real part of

$$V = \overline{V_F}e^{j(\omega t - \beta x)} + \overline{V_B}e^{j(\omega t + \beta x)} \tag{6.28}$$

In this case, $\overline{V_F}$ and $\overline{V_B}$ are complex numbers representing both the amplitude and phase offset of the voltage, where

$$\overline{V_F} = V_F e^{j\phi_F} \tag{6.29a}$$

$$\overline{V_B} = V_B e^{j\phi_B} \tag{6.29b}$$

Likewise, a solution to Eq. (6.15) for current is

$$I = I_F \cos(\omega t - \beta x + \phi_F) + I_B \cos(\omega t + \beta x + \phi_B) \tag{6.30}$$

where $I_F \cos(\omega t - \beta x + \phi_F)$ is the forward wave and $I_B \cos(\omega t + \beta x + \phi_B)$ is the backward wave. This can be expressed in complex notation as

$$I = \overline{I_F}e^{j(\omega t - \beta x)} + \overline{I_B}e^{j(\omega t + \beta x)} \tag{6.31}$$

where

$$\overline{I_F} = I_F e^{j\phi_F} \tag{6.32a}$$

$$\overline{I_B} = I_B e^{j\phi_B} \tag{6.32b}$$

As we shall see in the remainder of this chapter, we can use Eqs. (6.28) and (6.31) to quantitatively understand the behaviour of the current and voltage on transmission lines in a variety of circumstances.

6.4 Waves on Lossy Transmission Lines

In Sections 6.2 and 6.3, we have assumed that the transmission line is 'ideal' in the sense that there are no mechanisms by which energy is dissipated as the electromagnetic wave travels along the line. All of the energy which enters at one end arrives at the other. In practice, although some transmission lines may come close to this ideal for short lengths, in many cases this is not true – particularly at high frequencies, for reasons which will become clear when we look at electromagnetic waves in conducting media in Chapter 10. We call such non-ideal lines *lossy*. There are two loss mechanisms to consider, and we will need to amend our equivalent circuit for a small length of the ideal transmission line in Figure 6.2 to accommodate these.

The first, and perhaps most obvious, is that the conductors which form the transmission line will have a non-zero resistance. Therefore, current flow along the conductors will result in energy being dissipated. This resistance simply scales with length of the transmission line, so to accommodate this in the equivalent circuit for the short length δx we simply need to define a resistance per unit length of the conductors R. Therefore, the resistance of

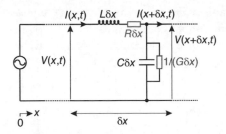

the short length of line is $R\delta x$. We might expect to have to attribute parts of this resistance to each of the conductors, but for a simple two-conductor system, it is mathematically equivalent to have a single resistance in series on one of the conductors which is the sum of the resistance of them both, as shown in Figure 6.9.

The second loss is associated with the dielectric between the conductors. We have assumed that this has an infinite resistance, so that there is no current flow between the conductors. In practice, the high electric fields which are present often result in a small parasitic current flow between the conductors which will cause energy dissipation in the dielectric. In this case, if we imagine the transmission line to be made up of many short lengths of transmission line, the dielectric resistances will all be in parallel with each other. As conductances in parallel simply sum, this is most simply incorporated into the equivalent circuit by defining a conductance per unit length G which appears between the two conductors. Therefore, the conductance is $G\delta x$ for the short length of line δx, as shown in Figure 6.9.

We can now analyse the lossy equivalent circuit in a similar fashion to that applied to the ideal case using the Kirchhoff laws. Once again, there is a potential difference between the two ends of the small length of line δV given by Eq. (6.6). In the ideal case, this voltage was dropped across the inductance term, but now some is dropped across the series resistance as well, and so Eq. (6.7) becomes

$$\delta V = -L\delta x\frac{dI}{dt} - IR\delta x \tag{6.33}$$

which can be rearranged in the limit as δV and δx tend to zero to

$$\frac{dV}{dx} = -L\frac{dI}{dt} - IR \tag{6.34}$$

Likewise, there is a difference in the current flow between the two ends of the small length of line δI defined by Eq. (6.9), which was the current through the capacitor in the ideal case, but is now the sum of the current through the capacitor and the conductance. Therefore, Eq. (6.10) becomes

$$-\delta I = C\delta x\frac{dV}{dt} + VG\delta x \tag{6.35}$$

which can be rearranged in the limit as δI and δx tend to zero to

$$\frac{dI}{dx} = -C\frac{dV}{dt} - VG \tag{6.36}$$

Equations (6.34) and (6.36) are the *telegrapher's equations* for a lossy transmission line. In the idealized case where both R and G tend to zero, they reduce back to the idealized equivalent (Eqs. (6.8) and (6.11)).

Just as with the idealized case, we can combine the telegrapher's equations to give differential equations for the voltage and current along the transmission line as a function of position and time by substituting one into the other. To get an expression for voltage, we start by differentiating Eq. (6.34) with respect to position:

$$\frac{d^2V}{dx^2} = -L\frac{d}{dx}\frac{dI}{dt} - R\frac{dI}{dx} \tag{6.37}$$

We can change the order of the first-order differentiation to give

$$\frac{d^2V}{dx^2} = -L\frac{d}{dt}\frac{dI}{dx} - R\frac{dI}{dx} \tag{6.38}$$

and substitute in the other telegrapher's equation (Eq. (6.36))

$$\frac{d^2V}{dx^2} = LC\frac{d^2V}{dt^2} + (LG + RC)\frac{dV}{dt} + RGV \tag{6.39}$$

to give a second-order differential equation that is only dependent on voltage, position and time. Alternatively, we can differentiate Eq. (6.36) with respect to position,

$$\frac{d^2I}{dx^2} = -C\frac{d}{dx}\frac{dV}{dt} - G\frac{dV}{dx} \tag{6.40}$$

and change the order of the first-order differentiation

$$\frac{d^2I}{dx^2} = -C\frac{d}{dt}\frac{dV}{dx} - G\frac{dV}{dx} \tag{6.41}$$

We can now substitute in the other telegrapher's equation (Eq. (6.34)) to give

$$\frac{d^2I}{dx^2} = LC\frac{d^2I}{dt^2} + (LG + RC)\frac{dI}{dt} + RGI \tag{6.42}$$

Comparing Eqs. (6.39) and (6.42) with their ideal equivalents (Eqs. (6.14) and (6.15)), it is clear that we have gained two new terms in each equation: a first-order differential term which is dependent on all of the equivalent circuit elements in Figure 6.9 and a simple non-differential term which is dependent only on the loss elements. Both equations simplify to their idealized equivalents if $R = 0$ and $G = 0$.

We need to seek a new solution to Eqs. (6.39) and (6.42). A solution to the idealized equations was a simple cosine wave, but this implies that no energy is being lost. Therefore, a sensible solution would be a decaying cosine wave of the form

$$\psi = \psi_0 \cos(\omega t - \beta x)e^{-\alpha x} \tag{6.43}$$

where the $e^{-\alpha x}$ term causes the wave to decay in amplitude with position as a result of the loss of energy in the series resistance and parallel conductance of the transmission line, as shown in Figure 6.10. The quantity α is often called the *attenuation coefficient*.

By way of an example, let us imagine that we have a forward-propagating wave only, for which the voltage can be expressed in complex notation (following that developed in Eq. (6.28)) as

$$V = \overline{V_F}e^{j(\omega t - \beta x)}e^{-\alpha x} \tag{6.44}$$

If we collect together position-dependent terms, then this can be rewritten as

$$V = \overline{V_F}e^{j\omega t}e^{-(\alpha + j\beta)x} = \overline{V_F}e^{j\omega t}e^{-\gamma x} \tag{6.45}$$

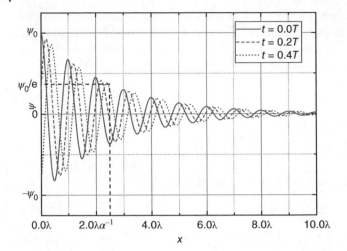

Figure 6.10 A wave of the form $\psi = \psi_0 \cos(\omega t - \beta x)e^{-\alpha x}$ plotted over a distance of ten wavelengths for three different times within the first time period T, clearly showing that the wave is moving in a positive x-direction and decaying in amplitude due to energy loss. The wave decays in amplitude by a factor of e over a distance of α^{-1}.

where γ is our new *propagation constant* for the lossy transmission line, given by

$$\gamma = \alpha + j\beta \tag{6.46}$$

We can gain insight into the physical meaning of the propagation constant by attempting to substitute Eq. (6.45) as a trial solution into the second-order differential equation for voltage on the lossy transmission line (Eq. (6.39)). The first-order differential of Eq. (6.45) with respect to position is

$$\frac{dV}{dx} = -(\alpha + j\beta)\overline{V_F}e^{j\omega t}e^{-(\alpha+j\beta)x} \tag{6.47}$$

and the second-order differential is

$$\frac{d^2V}{dx^2} = (\alpha + j\beta)^2\overline{V_F}e^{j\omega t}e^{-(\alpha+j\beta)x} \tag{6.48}$$

Meanwhile, the second-order differential of Eq. (6.45) with respect to time is

$$\frac{d^2V}{dt^2} = -\omega^2\overline{V_F}e^{j\omega t}e^{-(\alpha+j\beta)x} \tag{6.49}$$

Substituting Eqs. (6.47)–(6.49) into Eq. (6.39) gives

$$(\alpha + j\beta)^2 = -LC\omega^2 + j\omega(LG + RC) + RG \tag{6.50}$$

where a common term of $\overline{V_F}e^{j\omega t}e^{-(\alpha+j\beta)x}$ has been cancelled out. Taking the square root of Eq. (6.50) and rearranging gives

$$\gamma = \alpha + j\beta = \sqrt{(R + j\omega L)(G + j\omega C)} \tag{6.51}$$

This equation allows the propagation constant to be numerically evaluated for any lossy equivalent circuit. A key quantity which emerges from this is α^{-1}; this has units of length and it is the characteristic distance over which the amplitude of the wave decays by a factor of $e(=2.718)$, as shown in Figure 6.10. It allows a calculation of how long a transmission line can be before the signal becomes too small for useful detection.

6.5 Characteristic Impedance

By using an equivalent circuit for a small length of a transmission line, we have shown that it is possible to derive the telegrapher's equations, both for ideal transmission lines (Eqs. (6.8) and (6.11)) and for lossy transmission lines (Eqs. (6.34) and (6.36)). We have then used these to derive equations for how the current and voltage must depend on both position and time, and this resulted in the wave equations. In this analysis, we considered either current or voltage. We did not consider how current and voltage relate to each other. If we look at one of the telegrapher's equations for the ideal transmission line

$$\frac{dI}{dx} = -C\frac{dV}{dt} \tag{6.11}$$

then it is clear that a direct relationship between current and voltage does exist.

To understand this relationship further for the case of the ideal transmission line, let us substitute into Eq. (6.11) the general wave solutions for the current and voltage in complex notation

$$V = \overline{V_F}e^{j(\omega t - \beta x)} + \overline{V_B}e^{j(\omega t + \beta x)} \tag{6.28}$$

$$I = \overline{I_F}e^{j(\omega t - \beta x)} + \overline{I_B}e^{j(\omega t + \beta x)} \tag{6.31}$$

Differentiating Eq. (6.28) with respect to t and Eq. (6.31) with respect to x yields

$$\left(-\beta\overline{I_F}e^{-j\beta x} + \beta\overline{I_B}e^{j\beta x}\right)je^{j\omega t} = -C\left(\omega\overline{V_F}e^{-j\beta x} + \omega\overline{V_B}e^{j\beta x}\right)je^{j\omega t} \tag{6.52}$$

We can clearly cancel the common factor of $je^{j\omega t}$ from both sides of this equation. However, we should also remember that Eqs. (6.28) and (6.31) each have components representing a forward travelling wave (denoted by the subscript F) moving in the positive x-direction and a backward travelling wave (denoted by the subscript B) moving in the negative x-direction. As the forward wave has terms with the common factor $e^{-j\beta x}$ and the backward wave has terms with the common factor $e^{j\beta x}$, the two can be equated separately and the common factor eliminated in each case to give

$$\frac{\overline{V_F}}{\overline{I_F}} = \frac{\beta}{\omega C} \tag{6.52a}$$

$$\frac{\overline{V_B}}{-\overline{I_B}} = \frac{\beta}{\omega C} \tag{6.52b}$$

A simple relationship clearly does exist between the current and voltage for each of a forward and backward travelling wave, and the two are identical except for the minus sign on the $\overline{I_B}$ term in Eq. (6.52b). This occurs because of how the voltage and current are defined in the equivalent circuit, as shown in Figure 6.11. For both the forward and backward wave, the voltage is defined as the potential difference of the upper conductor with respect to the lower conductor, which does not depend on the direction of travel of the wave. However, the

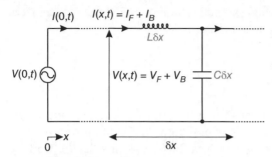

Figure 6.11 The definition of voltage on the transmission line is independent of direction of the wave as it is the potential difference of the upper conductor with respect to the lower conductor, whereas current is directional and changes sign depending on whether it is travelling forward from left to right or backward from right to left.

current travels along the conductors, and so has a direction, with a positive current being one travelling from left to right along the upper conductor. Therefore, the backward current in this general frame of reference must be the negative of the backward current in its own frame of reference.

Therefore, from Eq. (6.52), we can define a new quantity Z_0 called the *characteristic impedance* as the ratio of the voltage and current of a unidirectional wave (i.e. in the absence of any wave travelling in the opposite direction) at any point on a transmission line where

$$Z_0 = \frac{\overline{V_F}}{\overline{I_F}} = \frac{\overline{V_B}}{-\overline{I_B}} = \frac{\beta}{\omega C} \tag{6.53}$$

From Eq. (6.23) we know that the wave velocity $c = \omega/\beta$, while from Eq. (6.24) we have $c = 1/\sqrt{LC}$, and so substituting these into Eq. (6.53) gives a general expression for the characteristic impedance:

$$Z_0 = \sqrt{\frac{L}{C}} \tag{6.54}$$

Therefore, the characteristic impedance only depends on the inductance and capacitance per unit length of the transmission line. Note the similarity with the wave velocity (Eq. (6.24)) which was also only dependent on the same two quantities. By way of an example, we can substitute into Eq. (6.54) the expressions for L and C which we derived for the microstrip line in Eqs. (6.1) and (6.5)

$$Z_0 = \sqrt{\frac{\mu_0 \mu_r d}{w} \bigg/ \frac{\varepsilon_0 \varepsilon_r w}{d}} = \frac{d}{w}\sqrt{\frac{\mu_0 \mu_r}{\varepsilon_0 \varepsilon_r}} \tag{6.55}$$

This result is typical for all ideal transmission lines, namely that the characteristic impedance is dependent only on the permittivity and permeability of the dielectric between the conductors multiplied by some geometry factor g,

$$Z_0 = g\sqrt{\frac{\mu_0 \mu_r}{\varepsilon_0 \varepsilon_r}} \tag{6.56}$$

where for the microstrip line $g = d/w$ (the ratio of the distance between the conductors to the width of the line).

As the characteristic impedance is a ratio of voltage and current, it has units of ohms (Ω). However, this leads to a common misconception that it is therefore related to the energy being dissipated in the transmission line. This should immediately feel wrong as, by definition, the ideal transmission line does not dissipate energy. A little further thought explains why this impedance cannot dissipate energy. Normally, when we think of a simple electric circuit component, like a resistor, the impedance is the ratio of the potential difference across the component to the current through the component. The current is moving across the potential difference and therefore the energy of the electrons is changing and energy is dissipated (as heat in the resistor). However, inspection of Figure 6.11 shows that for the characteristic impedance, the voltage is the potential difference between the two conductors, but the current is flowing along one conductor – it does not flow directly from one conductor to the other across the potential difference. Therefore, no energy is being dissipated.

The characteristic impedance is purely a useful mathematical construction which relates the voltage to the current at a point in the transmission line at some moment in time. A good way of thinking about it is that if we were to connect an a.c. voltage source to one end of an infinitely long transmission line (so that there is only a unidirectional wave travelling in the forward direction) then the characteristic impedance would be the ratio of the voltage and current output by the voltage source. This is the situation shown in Figure 6.11, and so $Z_0 = V(0, t)/I(0, t)$.

It is also possible to derive a characteristic impedance for the lossy transmission line using Eq. (6.36) as the starting telegrapher's equation. We need to substitute into this an expression for generic forward and backward waves. This is simplest mathematically if the propagation constant is used (as in Eq. (6.45)). Therefore, the voltage wave can be expressed as

$$V = \overline{V_F} e^{j\omega t} e^{-\gamma x} + \overline{V_B} e^{j\omega t} e^{\gamma x} \qquad (6.57)$$

Note that it is the sign of the γx exponent that determines whether the wave is travelling in the positive x-direction (for $e^{-\gamma x}$) or negative x-direction (for $e^{\gamma x}$). Similarly, the current expression is

$$I = \overline{I_F} e^{j\omega t} e^{-\gamma x} + \overline{I_B} e^{j\omega t} e^{\gamma x} \qquad (6.58)$$

Substitution of Eqs. (6.57) and (6.58) into Eq. (6.36) gives

$$\gamma(-\overline{I_F} e^{-\gamma x} + \overline{I_B} e^{\gamma x}) = -j\omega C(\overline{V_F} e^{-\gamma x} + \overline{V_B} e^{\gamma x}) - G(\overline{V_F} e^{-\gamma x} + \overline{V_B} e^{\gamma x}) \qquad (6.59)$$

where a common factor of $e^{j\omega t}$ has been eliminated. This is analogous to Eq. (6.52); we can similarly separately equate the forward wave terms

$$\frac{\overline{V_F}}{\overline{I_F}} = \frac{\gamma}{G + j\omega C} \qquad (6.60)$$

and backward wave terms

$$\frac{\overline{V_B}}{-\overline{I_B}} = \frac{\gamma}{G + j\omega C} \qquad (6.61)$$

Therefore, using the same definition of the characteristic impedance as the ratio of voltage and current for a unidirectional wave at any point,

$$Z_0 = \frac{\overline{V_F}}{\overline{I_F}} = \frac{\overline{V_B}}{-\overline{I_B}} = \frac{\gamma}{G + j\omega C} \qquad (6.62)$$

We can now substitute for γ in terms of circuit component terms using Eq. (6.51) to give the general expression for the characteristic impedance of a lossy transmission line:

$$Z_0 = \sqrt{\frac{R + j\omega L}{G + j\omega C}} \tag{6.63}$$

Note that this reduces back to the expression for the ideal transmission line (Eq. (6.54)) if both $R = 0$ and $G = 0$, as expected. Further comparison with the ideal case shows that, while the characteristic impedance for the ideal line is entirely real, for the lossy line this is a complex number. Therefore, there is a phase difference between the voltage and current waves on the lossy line. Furthermore, the complex impedance has a frequency dependence for the lossy line which is also absent for the ideal line. Although the characteristic impedance of the lossy transmission line does contain terms associated with energy dissipation (R and G) it still cannot be used to calculate power dissipation as the voltage and current are not between the same points in space (i.e. the same situation as for the ideal transmission line).

6.6 Loads on Transmission Lines

The purpose of defining a characteristic impedance becomes clearer when we want to know how a transmission line will respond when other devices are connected to either end of the line or when transmission lines are connected to each other. In this section, we will consider specifically what happens when a load is connected to the end of a transmission line. In practice, this will almost always be the case. For example, we will probably want to connect a load to the end of a power transmission line, such as a motor, heating element or lighting, to transduce the electrical energy into some other form. Alternatively, we may have some analogue signal processing circuit at the coaxial cable input to a television from an aerial, or in a digital logic circuit the input to a logic gate will probably sit at the end of a microstrip line. In all of these cases, the load will have some input impedance Z_L.

To understand quantitatively how the voltage and current waves on the transmission line respond to the presence of such a load, let us consider the general situation shown in Figure 6.12 where we have a load impedance connected to the end of a transmission line at a point $x = 0$. We will imagine that the line extends infinitely into the negative x-direction so that we do not have to consider any effect of an impedance at the other end

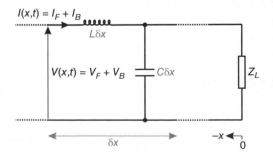

Figure 6.12 An infinitely long ideal transmission line with a load impedance Z_L at one end where $x = 0$.

of the line. For simplicity, we will use the ideal transmission line to understand how the system behaves, but the same result applies to lossy lines as well.

Let us imagine that we have a forward wave travelling in the positive x-direction on the transmission line. The load is analogous to having a mass attached to the end of a slinky spring with a forward longitudinal wave propagating along its length. Some of the kinetic and potential energy in the spring will be transmitted into the load mass, causing it to move, but some energy will be reflected along the spring as a backward travelling wave. The same is true in the transmission line; some of the energy will not be absorbed by the load but will be reflected as a backward travelling wave. Therefore, there will be both a forward and backward travelling wave on the line, and this can be represented by the general solutions that we derived previously:

$$V(x,t) = \overline{V_F}e^{j(\omega t - \beta x)} + \overline{V_B}e^{j(\omega t + \beta x)} \tag{6.28}$$

$$I(x,t) = \overline{I_F}e^{j(\omega t - \beta x)} + \overline{I_B}e^{j(\omega t + \beta x)} \tag{6.31}$$

We are specifically interested in what is happening at the load where $x = 0$, and so these simplify to

$$V(0,t) = (\overline{V_F} + \overline{V_B})e^{j\omega t} \tag{6.64a}$$

$$I(0,t) = (\overline{I_F} + \overline{I_B})e^{j\omega t} \tag{6.64b}$$

It should be noted that we would get the same equations if we used the more complicated expressions for V and I on the lossy transmission line (Eqs. (6.57) and (6.58)). These are the voltage and current across the load impedance, and so must also be related to Z_L by

$$Z_L = \frac{V(0,t)}{I(0,t)} = \frac{\overline{V_F} + \overline{V_B}}{\overline{I_F} + \overline{I_B}} \tag{6.65}$$

However, we also know that the forward voltage and current are related to each other by the characteristic impedance Z_0 of the transmission line, as are the backward voltage and current, according to Eq. (6.53). Therefore, we can substitute the current terms in Eq. (6.65) for voltage terms:

$$Z_L = \frac{\overline{V_F} + \overline{V_B}}{\overline{V_F}/Z_0 - \overline{V_B}/Z_0} \tag{6.66}$$

We can now rearrange this equation to find out how much of the forward wave is reflected as a backward wave by making $\overline{V_B}$ the subject:

$$\overline{V_B} = \overline{V_F}\left(\frac{Z_L - Z_0}{Z_L + Z_0}\right) \tag{6.67}$$

This allows us to define the *voltage reflection coefficient* as the proportion of the forward travelling wave on a transmission line that is reflected by a load as a backward travelling wave:

$$\rho_L = \frac{Z_L - Z_0}{Z_L + Z_0} \tag{6.68}$$

Examination of Eq. (6.68) suggests that there are three 'special cases' for particular values of load impedance Z_L terminating a transmission line with a characteristic impedance Z_0.

The first 'special case' is when the transmission line is terminated with an open circuit, so that the load impedance $Z_L = \infty$ (e.g. a coaxial cable with nothing attached to the end). In this case, the voltage reflection coefficient $\rho_L = 1$, which means that $\overline{V_B} = \overline{V_F}$. In other words, when the forward wave reaches the end of the transmission line, there is nowhere for the energy to go and so all of the wave must be reflected. Given that there is an open circuit, there is no constraint on voltage and so the backward wave is in phase with the forward wave at the reflection point ($x = 0$) and the two constructively interfere with each other, with the result that the amplitude of the total voltage at that point can be as large as $2|V_F|$. You may have noticed a similar effect in a swimming pool where the size of the waves at the walls of the pool seems to be much greater than elsewhere. This is because waves arriving at the wall have to be reflected, and with no constraint on the height of the water, the reflected wave is in phase with the incoming wave, giving an amplitude at the wall which is twice that of the incoming wave. We can see this mathematically as well. If $\overline{V_B} = \overline{V_F}$ then Eq. (6.28) becomes

$$V(x,t) = \overline{V_F} e^{j\omega t} (e^{j\beta x} + e^{-j\beta x}) \tag{6.69}$$

the real part of which can be re-expressed as

$$V(x,t) = 2\overline{V_F} \cos(\omega t) \cos(\beta x) \tag{6.70}$$

This is plotted in Figure 6.13 for different times. At $x = 0, -\lambda/2, -\lambda, \ldots$ the term $\cos(\beta x) = 1$ (remembering that $\beta = 2\pi/\lambda$). Therefore, at these points, the voltage oscillates with an amplitude of $2\overline{V_F}$ and an angular frequency of ω because the forward and backward waves are always in phase with each other. These points are called *antinodes*. However, at $x = -\lambda/4, -3\lambda/4, -5\lambda/4, \ldots$ the term $\cos(\beta x) = 0$ and therefore the voltage is always zero at these points because the forward and backward waves are always in antiphase with each other and perfectly cancel out. These points are called *nodes*. This characteristic pattern of a wave with nodes and antinodes is called a *standing wave* as the maxima of the total wave appear not to be moving spatially. Standing waves form when two waves of the same frequency and velocity travelling in opposite directions interfere with each other, which is the situation here. Therefore, for this special case where $Z_L = \infty$, we have a standing wave with an antinode at the reflection point.

The second 'special case' is when the transmission line is terminated with a closed circuit, so that the load impedance $Z_L = 0$ (e.g. a coaxial cable with a short circuit between the inner and outer conductors attached to the end). In this case, the voltage reflection coefficient $\rho_L = -1$, which means that $\overline{V_B} = -\overline{V_F}$. In other words, when the forward wave reaches the end of the transmission line, there is again nowhere for the energy to go and so all of the wave must be reflected, but this time there is a constraint. The short circuit means that there can be no potential difference between the two conductors at the load – the total voltage at this point must always be zero, so $V(0, t) = 0$. This causes the wave to undergo a π phase shift on reflection so that the forward and backward waves are in antiphase at this point. The sign of ρ_L indicates this phase shift. The result is that a node is formed at the reflection point $x = 0$ this time. At $x = -\lambda/4$, however, the forward and backward waves will be in phase,

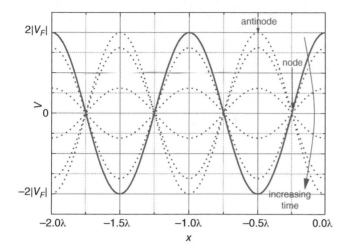

Figure 6.13 The sum for the forward and backward wave plotted for different times over half a time period where the transmission line is terminated with an infinite load impedance at $x = 0$. An antinode is formed at $x = 0$ and a node at $x = -\lambda/4$.

as the backward wave will have travelled an extra distance of $\lambda/2$ to get back to this point compared with the forward wave and will have undergone a π phase change on reflection to give a total phase change of 2π. An antinode will therefore be formed. We can see this mathematically as well. If $\overline{V_B} = -\overline{V_F}$ then Eq. (6.28) becomes

$$V(x,t) = \overline{V_F}e^{j\omega t}(e^{j\beta x} - e^{-j\beta x}) \tag{6.71}$$

the real part of which can be re-expressed as

$$V(x,t) = -2\overline{V_F}\sin(\omega t)\sin(\beta x) \tag{6.72}$$

This is plotted in Figure 6.14 for different times. At $x = 0, -\lambda/2, -\lambda, \dots$ the term $\sin(\beta x) = 0$ and therefore the voltage is always zero at these points because the forward and backward waves are always in antiphase with each other and perfectly cancel out to give nodes. At $x = -\lambda/4, -3\lambda/4, -5\lambda/4, \dots$ the term $\sin(\beta x) = 1$ and the voltage oscillates with an amplitude of $2\overline{V_F}$ because the forward and backward waves are always in phase with each other to give antinodes. Therefore, for this special case where $Z_L = 0$, we have a standing wave with a node at the reflection point.

In reality, most loads on the end of a transmission line will be somewhere between these two extreme 'special cases', and a proportion of the forward wave will be reflected. This situation is shown in Figure 6.15 for the situation where $\rho_L = 0.3$ and the forward, backward and total waves are all plotted over the first 0.3 of a time period T. The peak in the forward wave is clearly moving from left to right and the peak in the backward wave from right to left. The total wave also appears to be moving from left to right, which is to be expected as the forward wave is greater than the backward wave. However, if we plot just the summed wave over an entire cycle, as shown in Figure 6.16 for the same situation with $\rho_L = 0.3$, then we observe that there are clearly positions where the amplitude of the wave is greatest (antinodes) and least (nodes) – we still have a standing wave. A helpful quantity is the ratio

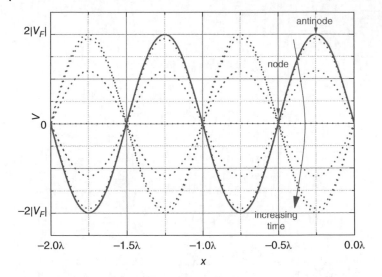

Figure 6.14 The sum for the forward and backward wave plotted for different times over half a time period where the transmission line is terminated with a zero load impedance at $x = 0$. A node is formed at $x = 0$ and an antinode at $x = -\lambda/4$.

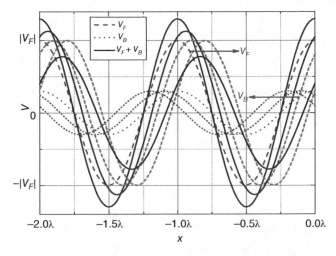

Figure 6.15 The forward (dashed), backward (dot) and total (solid) wave plotted at time $t = 0$ (heavy line) and $t = 0.1T$, $0.2T$ (light lines) for the situation where there is a load impedance at $x = 0$ giving a voltage reflection coefficient $\rho_L = 0.3$.

of the voltage amplitude at the antinodes to that at the nodes, which is commonly called the *voltage standing wave ratio* (VSWR), and is given by

$$\text{VSWR} = \frac{|\overline{V_F}| + |\overline{V_B}|}{|\overline{V_F}| - |\overline{V_B}|} = \frac{1 + |\rho_L|}{1 - |\rho_L|} \tag{6.73}$$

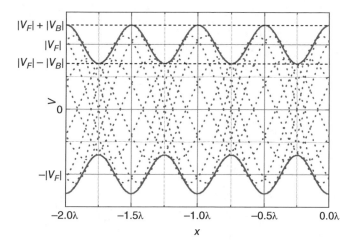

Figure 6.16 The total wave plotted at intervals of $0.1T$ over an entire period for the situation where there is a load impedance at $x = 0$ giving a voltage reflection coefficient $\rho_L = 0.3$.

For the two 'special cases' of infinite and zero load impedances, the VSWR $= \infty$, whereas for the case when $\rho_L = 0.3$, as in Figure 6.16, the VSWR $= 1.86$.

It is also notable in Figure 6.16 that there is still an antinode at the reflection point $(x = 0)$ as the voltage reflection coefficient is positive. This will clearly always be true from Eq. (6.68) if $Z_L > Z_0$. However, if $Z_L < Z_0$ then the voltage reflection coefficient will be negative and there will be a π phase change when the wave is reflected leading to a node at $x = 0$. This suggests that there is a third 'special case' which we have not yet considered, when $Z_L = Z_0$. In this condition, we can see from Eq. (6.68) that $\rho_L = 0$ and none of the wave is reflected. $\overline{V_B} = 0$ and all of the energy in the forward wave is dissipated in the load. In many situations this is the ideal outcome. For example, for the case of the coaxial cable transmitting signal from an aerial to a television receiver we want as much of the small signal power as possible to be dissipated in the receiving circuit and we certainly do not want a signal to be reflected back to the aerial which may interfere with the incoming signal (see Section 6.8 on ringing). Therefore, standards have been adopted for many cables and equipment that is attached to them. For example, most coaxial cables have characteristic impedances of either 50 or 75 Ω and most equipment with coaxial inputs or output has associated input or output impedances of 50 or 75 Ω as well to ensure maximum power transfer and minimum signal reflection.

6.7 Connections Between Transmission Lines

Throughout the previous section, we assumed that the load on the end of the transmission line is a simple component, such as a resistor. However, it may be that we wish to connect two transmission lines together, as shown in Figure 6.17. In the absence of any reflections on the second line (we will consider this case in more detail shortly) the characteristic impedance of the second line Z_{02} behaves just like the load impedance that we have

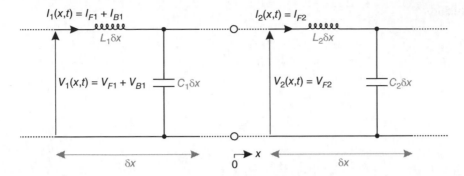

Figure 6.17 Two transmission lines are connected together at $x = 0$. A forward wave is being transmitted from the first transmission line on the left into the second transmission line on the right. Reflection may occur at the connection point resulting in a backward wave in the first line, but it is assumed that there is no backward wave in the second line.

already considered. Therefore, if the characteristic impedance of the first line is Z_{01}, using Eq. (6.68) we can see that the voltage reflection coefficient will be

$$\rho_L = \frac{Z_{02} - Z_{01}}{Z_{02} + Z_{01}} \tag{6.74}$$

If we want to ensure that there is no reflected wave backward along the first transmission line, we need to ensure that the two lines have the same characteristic impedance. However, there is a new quantity that we may wish to know – the proportion of the forward voltage wave $|\overline{V_{F1}}|$ in the first transmission line which is transmitted into the second transmission line as a forward voltage wave $|\overline{V_{F2}}|$. We call this ratio the *voltage transmission coefficient*, and it is given by

$$\rho_T = \frac{|\overline{V_{F2}}|}{|\overline{V_{F1}}|} \tag{6.75}$$

It is tempting to think that this could be simply calculated by ensuring that the transmission and reflection coefficients sum to unity, but this is not correct as it is the tangential component of the electric field **E** that is conserved at the dielectric interface between the two transmission lines and not the voltage (see Section 9.2). However, we can use the fact that the current is conserved at the interface between the two transmission lines, which means that

$$|\overline{I_{F2}}| = |\overline{I_{F1}}| + |\overline{I_{B1}}| \tag{6.76}$$

where we are allowing for a backward wave in the first transmission line due to reflection at the connection to the second transmission line, but we are assuming that there are no reflections in the second line (either because it is infinitely long or it is terminated with a load to give no reflections). We can rewrite Eq. (6.75) in terms of currents using the characteristic impedances of the two transmission lines to give

$$\rho_T = \frac{|\overline{V_{F2}}|}{|\overline{V_{F1}}|} = \frac{Z_{02}|\overline{I_{F2}}|}{Z_{01}|\overline{I_{F1}}|} \tag{6.77}$$

By substituting Eq. (6.76) into Eq. (6.77) we have

$$\rho_T = \frac{|\overline{V_{F2}}|}{|\overline{V_{F1}}|} = \frac{Z_{02}(|\overline{I_{F1}}| + |\overline{I_{B1}}|)}{Z_{01}|\overline{I_{F1}}|} = \frac{Z_{02}}{Z_{01}} \left(1 + \frac{|\overline{I_{B1}}|}{|\overline{I_{F1}}|} \right) \tag{6.78}$$

We can now relate this back to voltages again using Eq. (6.53),

$$\rho_T = \frac{Z_{02}}{Z_{01}} \left(1 - \frac{|\overline{V_{B1}}|}{|\overline{V_{F1}}|} \right) \tag{6.79}$$

but $|\overline{V_{B1}}|/|\overline{V_{F1}}|$ is just the voltage reflection coefficient, so substitution of Eq. (6.74) gives

$$\rho_T = \frac{Z_{02}}{Z_{01}}(1 - \rho_L) = \frac{Z_{02}}{Z_{01}} \left(1 - \frac{Z_{02} - Z_{01}}{Z_{02} + Z_{01}} \right) \tag{6.80}$$

which finally rearranges as

$$\rho_T = \frac{2Z_{02}}{Z_{02} + Z_{01}} \tag{6.81}$$

This equation allows the proportion of the voltage wave that is transmitted from one transmission line into another to be quickly calculated from the characteristic impedances. It can be seen that $\rho_T = 1$ for the special case of $Z_{02} = Z_{01}$ when there is no reflection, as would be expected. We should also note that, as the derivation of the voltage transmission coefficient is only dependent on what is happening at the connection point, the result is the same for both lossy and ideal transmission lines.

6.8 Input Impedance of an Ideal Terminated Line

In Section 6.6, we saw that reflections will occur at the end of a loaded transmission line which will result in a backward wave propagating along the line back towards any input which was creating the source signal in the first place. This creates a practical problem. The characteristic impedance of a transmission line is defined as the ratio of the voltage and current of a unidirectional wave (i.e. in the absence of any wave travelling in the opposite direction) at any point on a transmission line (see Section 6.5). In the presence of both a forward and backward wave, the actual ratio of total current and total voltage at any point on the line will not be the same as the characteristic impedance. In considering connecting two transmission lines together in Section 6.7, we got around this problem by assuming that there were no reflections on the second transmission line, and so the apparent impedance at the input was just the characteristic impedance. However, we now need to be able to work out the apparent input impedance of a transmission line of finite length which is terminated with a load impedance in order to be able to tackle practical problems where there are things connected to both ends of the transmission line.

To this end, let us consider the situation shown in Figure 6.18 where we have an ideal transmission line of characteristic impedance Z_0 which is terminated with a load impedance at position $x = 0$. The transmission line has a finite length l, and so the position of the input is at $x = -l$.

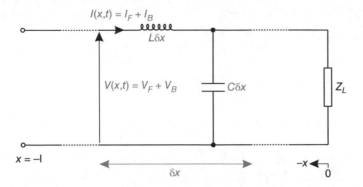

Figure 6.18 A transmission line of finite length *l* which is terminated with a load impedance Z_L at $x = 0$.

We can calculate the apparent impedance (the ratio of the total voltage to total current) at any point along the line using Eqs. (6.28) and (6.31):

$$Z(x) = \frac{V(x, t)}{I(x, t)} = \frac{\overline{V_F}e^{j(\omega t - \beta x)} + \overline{V_B}e^{j(\omega t + \beta x)}}{\overline{I_F}e^{j(\omega t - \beta x)} + \overline{I_B}e^{j(\omega t + \beta x)}} \tag{6.82}$$

The first thing to note is that a common factor of $e^{j\omega t}$ cancels out, and so the apparent impedance at any point is time-independent, which is an important result in itself. We can then use the definition of the characteristic impedance in terms of forward and backward currents and voltages (Eq. (6.53)) to eliminate current from Eq. (6.82):

$$Z(x) = \frac{\overline{V_F}e^{-j\beta x} + \overline{V_B}e^{j\beta x}}{\frac{\overline{V_F}}{Z_0}e^{-j\beta x} - \frac{\overline{V_B}}{Z_0}e^{j\beta x}} \tag{6.83}$$

Collecting together Z_0 and dividing all terms by $\overline{V_F}$ then gives

$$Z(x) = Z_0 \left(\frac{e^{-j\beta x} + \frac{\overline{V_B}}{\overline{V_F}}e^{j\beta x}}{e^{-j\beta x} - \frac{\overline{V_B}}{\overline{V_F}}e^{j\beta x}} \right) \tag{6.84}$$

which allows us to re-express the apparent impedance in terms of the voltage reflection coefficient using its definition as the ratio of the backward and forward voltages:

$$Z(x) = Z_0 \left(\frac{e^{-j\beta x} + \rho_L e^{j\beta x}}{e^{-j\beta x} - \rho_L e^{j\beta x}} \right) \tag{6.85}$$

We can also show explicitly how the apparent impedance depends on the load impedance using Eq. (6.68) to substitute for ρ_L:

$$Z(x) = Z_0 \left[\frac{(Z_L + Z_0)e^{-j\beta x} + (Z_L - Z_0)e^{j\beta x}}{(Z_L + Z_0)e^{-j\beta x} - (Z_L - Z_0)e^{j\beta x}} \right] \tag{6.86}$$

It is clear that the apparent impedance is varying along the line, but it is perhaps less obvious how it is varying as a result of the complex exponentials. Therefore, it is helpful to

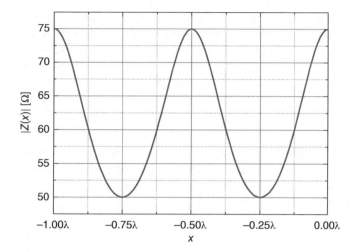

Figure 6.19 The apparent impedance along a 61.24 Ω characteristic impedance transmission line with a termination load of 75 Ω at $x = 0$.

use the de Moivre theorem to express Eq. (6.86) in terms of trigonometric functions:

$$Z(x) = Z_0 \left[\frac{Z_L - jZ_0 \tan(\beta x)}{Z_0 - jZ_L \tan(\beta x)} \right] \tag{6.87}$$

By way of an example, Figure 6.19 uses this equation to plot the magnitude of the apparent impedance as a function of position along a transmission line with a characteristic impedance of 61.24 Ω over a distance of a wavelength from a 75 Ω load. At the load ($x = 0$), Eq. (6.87) reduced to just Z_L. This should be expected as it is the load defining the impedance at this point, and this can be seen in Figure 6.19 as the impedance is 75 Ω at $x = 0$. The apparent impedance is then a smoothly varying function away from this point, oscillating over a distance of $\lambda/2$.

For a transmission line of length l we can evaluate Eq. (6.87) at $x = -l$ to work out an explicit expression for the input impedance Z_i of the transmission line which would effectively be 'seen' by any device connected to the input:

$$Z_i = Z_0 \left[\frac{Z_L + jZ_0 \tan(\beta l)}{Z_0 + jZ_L \tan(\beta l)} \right] \tag{6.88}$$

Close inspection of this equation shows that there is a 'special case' when $l = \lambda/4$, as at this point $\beta l = \pi/2$ and $\tan(\beta l) = \infty$, meaning that Eq. (6.88) reduces to

$$Z_i = \frac{Z_0^2}{Z_L} \tag{6.89}$$

Therefore, if we connect a device which has an output impedance of Z_i or another transmission line with a characteristic impedance of Z_i to the input of this transmission line, there will be no reflections, even though the load impedance at the end of the system is completely different. We have managed to connect two devices with completely different impedances without any reflections. This is known as *quarter-wave matching*.

For example, let us say that we wish to connect a transmission line with a characteristic impedance of $50\,\Omega$ to a load with an impedance of $75\,\Omega$. We can achieve this by first connecting another transmission line with a quarter-wavelength length to the load. We know that we want this short length of transmission line to have an input impedance $Z_i = 50\,\Omega$, and so its characteristic impedance according to Eq. (6.89) needs to be

$$Z_0 = \sqrt{Z_i Z_L} = \sqrt{50 \times 75} = 61.24\,\Omega \tag{6.90}$$

This is the situation shown at $x = -\lambda/4$ in Figure 6.19. Reflections are being avoided because they only occur when there is a step-change in impedance. However, in this situation we have created a system where the input impedance to the quarter wavelength of transmission line is $50\,\Omega$, and so connecting a $50\,\Omega$ characteristic impedance transmission line to this input will not cause an incoming wave to suffer any reflections. As the wave propagates along the quarter-wavelength line, it 'sees' a smoothly varying impedance increasing up to $75\,\Omega$ when it reaches the $75\,\Omega$ load, and so there is no reflection at this point either. All the input energy is dissipated in the load impedance, as required. This is an excellent example of how a quantitative understanding of how transmission lines work allows us to engineer optimum solutions to specific problems.

In this section, we have only considered ideal transmission lines. This is because, in practice, if the transmission line is lossy, then, by minimizing reflections at the load, the natural attenuation of a backward wave as it travels along the line back towards the input means that it becomes insignificant. Therefore, in general we only need to worry about the input impedance being significantly different from the characteristic impedance for low loss lines or lines which are short compared with the attenuation distance α^{-1}.

6.9   Ringing

Thus far in this chapter, we have concerned ourselves only with analysing what happens when we send a continuous, steady signal along a transmission line. However, in many cases, we will wish the signal to be time-varying, and therefore the transient response of the line can become significant. This is particularly true in digital communications, where we are typically sending data encoded as low and high voltages to represent 0 and 1 in binary. Therefore, we will be trying to send step-changes in voltage along a transmission line (rather like a square-wave signal). Ideally, we would want the voltage along the transmission line to change instantaneously in response to a step-change in voltage at the input, but in practice it takes some time for a response to occur, which we call *ringing*.

To carry out a basic analysis of this, let us consider the circuit shown in Figure 6.20, where we have a simple d.c. voltage source V_i with an output impedance Z_i connected via a simple switch to a transmission line, where a digital signal can be sent down the line by opening and closing the switch. The transmission line is ideal and has a length l and a characteristic impedance Z_0. At the other end of the line, at position $x = 0$, a simple load impedance Z_L has been connected. We will assume that the switch has been open for a long time so that the voltage on the transmission line is zero everywhere. Let the switch close at time $t = 0$. We will now work out how the voltage on the line responds to the resulting step-change in applied voltage at the source.

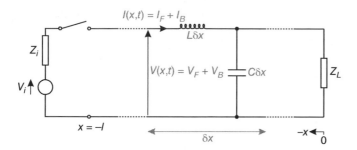

Figure 6.20 A transmission line of finite length l which is terminated with a load impedance Z_L at $x = 0$ and with an input dc voltage source V_i with an output impedance Z_i connected to the input at $x = -l$.

We know that a signal can only travel down the transmission line at a finite velocity given by Eq. (6.24). Therefore, at the moment the switch is closed, the load can have no effect on the voltage that is input onto the transmission line. The voltage source effectively 'sees' the transmission line as though it was a simple load impedance of Z_0, and so there is a potential division of the voltage V_i across Z_i and Z_0, and so the voltage actually appearing across the input end of the line at $x = -l$ and $t = 0$ is simply

$$V(x = -l, t = 0) = V_i \left(\frac{Z_0}{Z_i + Z_0} \right) \tag{6.91}$$

This voltage will then be transmitted along the line at a velocity $c = 1/\sqrt{LC}$ (from Eq. (6.24)), arriving at the load some time T later, where

$$T = \frac{l}{c} = l\sqrt{LC} \tag{6.92}$$

On reaching the load, some of the input voltage will be reflected, as given by the reflection coefficient

$$\rho_L = \frac{Z_L - Z_0}{Z_L + Z_0} \tag{6.68}$$

and this will sum with the input voltage to give the total voltage at the load ($x = 0$) at time $t = T$:

$$V(x = 0, t = T) = V_i \left(\frac{Z_0}{Z_i + Z_o} \right) [1 + \rho_L] \tag{6.93}$$

This voltage signal will take a further time T to travel back to the input end of the transmission line, where the output impedance of the voltage source will now look like a load, causing reflection of a proportion of the voltage back along the transmission line again. From Eq. (6.68), this voltage reflection coefficient at the input end of the line ρ_i will be

$$\rho_i = \frac{Z_i - Z_0}{Z_i + Z_0} \tag{6.94}$$

This will also sum with the reflected voltage, so the total voltage at the input end of the line after a complete round-trip of the signal will be

$$(x = -l, t = 2T) = V_i \left(\frac{Z_0}{Z_i + Z_o} \right) [1 + \rho_L + \rho_L \rho_i] \tag{6.95}$$

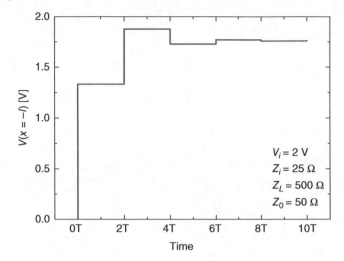

Figure 6.21 Response with time at the input end of a 50 Ω transmission line with a 2 V d.c. source with an output impedance of 25 Ω connected at the input and a load impedance of 50 Ω connected at the output.

If there is another complete round-trip of the signal, taking a further time $2T$, then the new voltage at the input will be

$$(x = -l, t = 4T) = V_i \left(\frac{Z_0}{Z_i + Z_o} \right) \left[1 + \rho_L + \rho_L \rho_i + \rho_L^2 \rho_i + \rho_L^2 \rho_i^2 \right] \tag{6.96}$$

We can now generalize this to a geometric series for the general case of there being n round-trips:

$$(x = -l, t = 2nT) = V_i \left(\frac{Z_0}{Z_i + Z_o} \right) \left[1 + \sum_{n=1}^{n} \left(\rho_L^n \rho_i^{n-1} + \rho_L^n \rho_i^n \right) \right] \tag{6.97}$$

As this is a geometric series, this will converge to a steady voltage with time, as shown for the example in Figure 6.21. Therefore, Eq. (6.97) allows an estimation of how long it will take the voltage on the transmission line to settle to an acceptable level.

Reference

Stroud, K.A. (1996). *Further Engineering Mathematics: Programmes and Problems*. Basingstoke: Macmillan.

7

Electromagnetic Waves in Dielectric Media

7.1 Introduction

In Chapter 5, we saw that there are four equations, collectively known as the Maxwell equations, which each individually say something about the origin and nature of electric and magnetic fields. These were summarized in Table 5.1 in both their integral and differential (vector calculus) forms, together with a summary of their physical interpretation. In this chapter we shall see that, by using them collectively, we can understand how electromagnetic waves behave in dielectrics. Such an understanding is of critical importance to engineers because we have learned to use electromagnetic waves across a broad range of frequencies (and therefore wavelengths) for a diversity of applications.

We have used the low-frequency end of the electromagnetic spectrum (from ~8 kHz up to ~300 GHz) for wireless communications since the early work on radio transmission by Guglielmo Marconi in the late nineteenth and early twentieth centuries. This led to television transmission in the 1920s. More recently, we have seen the rise of wireless communications, including medium-range mobile phone communications with ever-increasing data rates and short-range communications based on Wi-Fi™, Bluetooth®, and other similar platforms. Life has been transformed by such wireless-based communications systems since the late 1990s. Today, the radio spectrum is heavily used for a diversity of applications from terrestrial digital television to satellite broadcasting, 4G mobile and space science, to name just a few. Within this part of the spectrum, we also find microwaves, which as well as being used for communications are also used for cooking, as frequencies ~2.45 GHz match the resonant frequencies of water molecules, allowing electromagnetic energy to be efficiently absorbed, resulting in heating.

The part of the electromagnetic spectrum around 1–30 THz has been historically little used as it is difficult to generate electromagnetic waves at these frequencies. However, as our ability to create terahertz emitters has improved, so we have started to find uses for electromagnetic waves of these frequencies, particularly for imaging.

In the region from 30 to 400 THz we find the infrared part of the electromagnetic spectrum which is widely used for optical fibre-based communications, before reaching visible light in the range of 400–800 THz with its wide diversity of applications, from displays to laser-based surgery. Just beyond the visible spectrum, we reach ultraviolet frequencies from 800 to 30 000 THz which have found applications including water sterilization and the patterning of semiconductor devices by photolithography.

At the high-frequency end of the spectrum, we find X-rays (10^{17}–10^{19} Hz) with their application in medical imaging, and finally gamma rays $\sim 10^{20}$ Hz which also have medical applications in radiotherapy treatment of cancer.

It is clear that engineers need a good appreciation of how electromagnetic waves interact with a range of materials, but we should start with the most basic of material systems: dielectrics.

7.2 The Maxwell Equations in Dielectrics

The Maxwell equations, which were described in detail in Chapter 5, apply to all media, and can be summarized in differential form as

$$\nabla . \mathbf{D} = \rho \tag{5.7}$$

$$\nabla . \mathbf{B} = 0 \tag{5.9}$$

$$\nabla \times \mathbf{E} = -\frac{\partial \mathbf{B}}{\partial t} \tag{5.17}$$

$$\nabla \times \mathbf{H} = \mathbf{J} + \frac{\partial \mathbf{D}}{\partial t} \tag{5.32}$$

Table 5.1 also gives the integral form of these equations and a brief summary of their physical interpretations.

In Section 2.2, we saw that dielectrics are materials which have no free charges in their bulk, with the consequence that they are very poor conductors of current flow. Vacuum behaves as a dielectric medium, as do gases including air (noting that ionized gases which do conduct are called plasmas and are considered to be a different state of matter with very distinct physics). Many liquids such as deionized water also behave as dielectrics, as do some solids including most plastics and glass.

As dielectrics have no free charge, there can be no flow of current in these materials, and therefore current density $\mathbf{J} = 0$ everywhere in such a material. Furthermore, the application of an electric field can only result in a surface polarization charge on a dielectric with no net charge density in the bulk (i.e. $\rho = 0$). These two properties of dielectrics mean that the Gauss law of electric fields (Eq. (5.7)) and the Ampère–Maxwell law (Eq. (5.32)) can be immediately simplified. Therefore, the four Maxwell equations can be rewritten for dielectric media as

$$\nabla . \mathbf{D} = 0 \tag{7.1a}$$

$$\nabla . \mathbf{B} = 0 \tag{7.1b}$$

$$\nabla \times \mathbf{E} = -\frac{\partial \mathbf{B}}{\partial t} \tag{7.1c}$$

$$\nabla \times \mathbf{H} = \frac{\partial \mathbf{D}}{\partial t} \tag{7.1d}$$

7.3   The Electromagnetic Wave Equation

We can now combine the Maxwell equations for dielectric media (Eq. (7.1)) to produce a wave equation for either electric or magnetic fields in dielectrics. To do this, we need to take the curl of the Faraday law of magnetic fields (Eq. (7.1c)) to give

$$\nabla \times \nabla \times \mathbf{E} = -\frac{\partial}{\partial t}(\nabla \times \mathbf{B}) \tag{7.2}$$

One of the common vector calculus identities is that for an arbitrary vector quantity $\mathbf{Y}$,

$$\nabla \times \nabla \times \mathbf{Y} = \nabla(\nabla . \mathbf{Y}) - \nabla^2 \mathbf{Y} \tag{7.3}$$

and so Eq. (7.2) can be rewritten as

$$\nabla(\nabla . \mathbf{E}) - \nabla^2 \mathbf{E} = -\frac{\partial}{\partial t}(\nabla \times \mathbf{B}) \tag{7.4}$$

However, in a dielectric medium where there is no net charge in the bulk of the material, we know that the Gauss law of electric fields (Eq. (7.1a)) simplifies to $\nabla . \mathbf{D} = 0$, and as $\mathbf{D} = \varepsilon_0 \varepsilon_r \mathbf{E}$ (Eq. (2.15)) then $\nabla . \mathbf{E} = 0$ also. Hence, Eq. (7.4) reduces to

$$\nabla^2 \mathbf{E} = \frac{\partial}{\partial t}(\nabla \times \mathbf{B}) \tag{7.5}$$

In general, dielectrics behave as simple ideal linear magnetic materials (see Section 4.2) in which $\mathbf{B} = \mu_0 \mu_r \mathbf{H}$ (Eq. (4.9)). Therefore, Eq. (7.5) becomes

$$\nabla^2 \mathbf{E} = \mu_0 \mu_r \frac{\partial}{\partial t}(\nabla \times \mathbf{H})$$

We can now substitute in the Ampère–Maxwell equation for dielectric media (Eq. (7.1d)) to give

$$\nabla^2 \mathbf{E} = \mu_0 \mu_r \frac{\partial^2 \mathbf{D}}{\partial t^2} \tag{7.6}$$

Finally, with the same substitution of Eq. (2.15) to convert electric flux density into electric field, we obtain the key result

$$\nabla^2 \mathbf{E} = \varepsilon_0 \varepsilon_r \mu_0 \mu_r \frac{\partial^2 \mathbf{E}}{\partial t^2} \tag{7.7}$$

This is a three-dimensional *wave equation*. We have previously seen examples of one-dimensional wave equations when considering transmission lines in Section 6.3. They are characterized by having the second differential of some quantity with respect to position proportional to the second differential of the same quantity with respect to time. Therefore, for some arbitrary vector quantity $\mathbf{Y}$, this can be expressed as

$$\nabla^2 \mathbf{Y} = \frac{1}{c^2} \frac{\partial^2 \mathbf{Y}}{\partial t^2} \tag{7.8}$$

where c is the wave velocity (as was previously shown in Section 6.3).

The derivation of Eq. (7.7) is of the utmost significance: it proves that electromagnetic waves are a direct consequence of the Faraday law and Ampère–Maxwell equation in a medium where there is no net charge. At one level, perhaps this is not surprising, as the Faraday law says that a changing magnetic field will produce a changing, circulating electric field and the Ampère–Maxwell equation says that a changing electric field will produce a changing, circulating magnetic field. Therefore, having initiated a changing electric or magnetic field, the Maxwell equations suggest that this should lead to a self-sustaining electromagnetic wave. However, in addition to this, quantitative comparison of Eqs. (7.7) and (7.8) reveals that the velocity of the wave must be

$$c = \frac{1}{\sqrt{\varepsilon_0 \varepsilon_r \mu_0 \mu_r}} \tag{7.9}$$

It is notable that this expression for the velocity of an electromagnetic wave is the same as that obtained for the velocity of a wave on a transmission line (Eq. (6.26)). This is because the conductors in the transmission line are simply guiding an electromagnetic wave along a path. The wave itself is in the volume of space around the conductors, which is a dielectric medium. Therefore, we should expect the two equations to be the same from a physical perspective. We would also expect this mathematically, as Eq. (6.26) is obtained from the inductance and capacitance per unit length of the transmission line, which themselves are derived from the Gauss law of electric fields (see Section 2.3) and the Ampère circuital law (see Section 3.3). In both cases, we have started with Maxwell equations, but we have gone via different routes to end at the same result.

We can also derive a wave equation for the magnetic field component of electromagnetic waves that has a similar form to Eq. (7.7) starting by taking the curl of the Ampère–Maxwell equation in dielectric media (Eq. (7.1d)):

$$\nabla \times \nabla \times \mathbf{H} = \frac{\partial}{\partial t}(\nabla \times \mathbf{D}) \tag{7.10}$$

Using the vector identity in Eq. (7.3), this can be rewritten as

$$\nabla(\nabla.\mathbf{H}) - \nabla^2 \mathbf{H} = \frac{\partial}{\partial t}(\nabla \times \mathbf{D}) \tag{7.11}$$

We also know from the Gauss law of magnetic fields (Eq. (7.1b)) that $\nabla.\mathbf{B} = 0$, and as dielectrics are simple ideal linear magnetic materials ($\mathbf{B} = \mu_0 \mu_r \mathbf{H}$) it follows that $\nabla.\mathbf{H} = 0$. Furthermore, as $\mathbf{D} = \varepsilon_0 \varepsilon_r \mathbf{E}$, Eq. (7.11) can be simplified to

$$-\nabla^2 \mathbf{H} = \varepsilon_0 \varepsilon_r \frac{\partial}{\partial t}(\nabla \times \mathbf{E}) \tag{7.12}$$

Finally, we can substitute in the Faraday law and again use $\mathbf{B} = \mu_0 \mu_r \mathbf{H}$ to give the wave equation expressed in terms of the magnetic field:

$$\nabla^2 \mathbf{H} = \varepsilon_0 \varepsilon_r \mu_0 \mu_r \frac{\partial^2 \mathbf{H}}{\partial t^2} \tag{7.13}$$

Comparison of Eqs. (7.7), (7.13), and (7.8) shows that the velocity of the electric and magnetic field components of an electromagnetic wave is therefore the same.

7.4 Refractive Index and Dispersion

It is apparent from Eq. (7.9) that the velocity of an electromagnetic wave is dependent on the medium through which it is passing. The velocity of an electromagnetic wave in a vacuum where $\varepsilon_r = 1$ and $\mu_r = 1$ is simply

$$c_{\text{vac}} = \frac{1}{\sqrt{\varepsilon_0 \mu_0}} \tag{7.14}$$

which is evaluated to be $2.998{\times}10^8$ m s^{-1} using the known values for ε_0 and μ_0. This proof of the velocity of light is one of the great achievements of the Maxell equations. The velocity of an electromagnetic wave in any other dielectric medium can then be expressed in terms of the velocity in a vacuum by

$$c = \frac{c_{\text{vac}}}{\sqrt{\varepsilon_r \mu_r}} \tag{7.15}$$

This allows a quantity n called the *refractive index* of a dielectric medium to be defined as

$$n = \frac{c_{\text{vac}}}{c} \tag{7.16}$$

As electromagnetic waves travel more slowly in most dielectric media compared with vacuum, the refractive index is almost always greater than 1, and a larger value indicates a slower wave velocity.

Substitution of Eqs. (7.9) and (7.14) into Eq. (7.16) allows the refractive index to be expressed in terms of the relative permittivity and relative permeability of the dielectric,

$$n = \sqrt{\varepsilon_r \mu_r} \tag{7.17}$$

and in practice, as most dielectrics have a relative permeability of 1,

$$n = \sqrt{\varepsilon_r} \tag{7.18}$$

What is not immediately obvious is that the relative permittivity of a dielectric, and therefore the refractive index, is frequency-dependent. When we considered the physics of dielectric polarization in Section 2.2, we imagined the application of a static (i.e. not time-varying) electric field applied to a material which led to a polarization. However, it is clear that this polarization process must take some finite time. In addition, it is possible that electromagnetic waves at certain frequencies can excite other processes. For example, if the frequency of the electromagnetic wave matches the resonant frequency of a molecule in the dielectric, which is particularly common in the infrared part of the spectrum, then a strong interaction will result. We will return to this in Chapter 10 when we consider in more detail the interaction of electromagnetic waves with materials that significantly absorb the energy in the wave.

For now, the important practical consequence of this frequency dependence of the refractive index is that electromagnetic waves of different frequencies passing through the same material will have different velocities. We call this phenomenon *dispersion*, and the manifestation of this that we are perhaps most familiar with is the splitting of a beam of visible white light into a rainbow of components by a prism. To understand the importance of dispersion, let us imagine that we have two arbitrary waves ψ_1 and ψ_2 in one-dimensional space travelling in a positive x-direction where

$$\psi_1 = \psi_0 \cos(\omega_1 t - \beta_1 x) \tag{7.19a}$$

$$\psi_2 = \psi_0 \cos(\omega_2 t - \beta_2 x) \tag{7.19b}$$

so that each wave has the same amplitude ψ_0, but different angular frequencies and propagation constants. Individually, we know from Eq. (6.23) that each wave will have a velocity given by $c = \omega/\beta$, which we should strictly call the *phase velocity*. However, the sum of the two waves is

$$\psi = \psi_1 + \psi_2 = \psi_0[\cos(\omega_1 t - \beta_1 x) + \cos(\omega_2 t - \beta_2 x)] \tag{7.20}$$

which, using the standard trigonometric identity

$$\cos\theta_1 + \cos\theta_2 = 2\cos\left(\frac{\theta_1 + \theta_2}{2}\right)\cos\left(\frac{\theta_1 - \theta_2}{2}\right) \tag{7.21}$$

becomes

$$\psi = 2\psi_0 \cos\left[\left(\frac{\omega_1 + \omega_2}{2}\right)t - \left(\frac{\beta_1 + \beta_2}{2}\right)x\right]\cos\left[\left(\frac{\omega_1 - \omega_2}{2}\right)t - \left(\frac{\beta_1 - \beta_2}{2}\right)x\right] \tag{7.22}$$

For the case where the two waves are of very similar frequencies, this shows that the resulting superposition is a wave with a frequency that is the average of the two (the first cosine term in Eq. (7.22)) which is modulated at a low frequency which is equal to half the difference between the frequency of the two individual waves (the second cosine term in Eq. (7.22)). An example of this is shown in Figure 7.1. This low-frequency modulation means that it looks as though we have a sequence of *packets* of waves. These packets appear to move at a different velocity than the individual waves, and we call this the *group velocity*. As it is the second of the cosine terms in Eq. (7.22) that defines the low-frequency modulation and therefore the group velocity v_g, we can see from Eq. (6.23) that

$$v_g = \frac{\omega_1 - \omega_2}{\beta_1 - \beta_2} \tag{7.23}$$

which in the limit for a small difference in frequency becomes

$$v_g = \frac{d\omega}{d\beta} \tag{7.24}$$

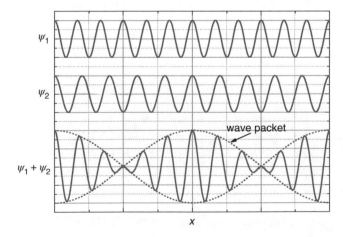

Figure 7.1 The two individual waves ψ_1 and ψ_2 from Eq. (7.19) each travel with slightly different phase velocities in a dispersive medium because they have different frequencies. The sum of the two is a wave that is modulated by a low-frequency cosine wave (indicated as the dashed line) into packets whose envelope travels at the group velocity.

It is this group velocity that actually determines the velocity with which either energy or information is being transmitted, and therefore it is this velocity that can never be greater than the velocity of light in a vacuum according to Einstein's special theory of relativity. It is for this reason that some materials can have a refractive index that is less than 1, which would give a phase velocity greater than the speed of light in a vacuum, as a calculation of the group velocity in those circumstances will always result in a value that is less than this critical velocity. For a material that is non-dispersive or only very weakly dispersive over a frequency range of interest (so that $\omega \propto \beta$), the group velocity will approximately equal the phase velocity.

7.5 Properties of Monochromatic Plane Electromagnetic Waves

In Section 7.3, we saw that the Maxwell equations can be combined to prove that electromagnetic waves can propagate through dielectric media, as a wave equation can be produced for both the electric field (Eq. (7.7)) and magnetic field (Eq. (7.13)). Furthermore, the electromagnetic wave has a characteristic velocity given by Eq. (7.9). We shall now study the other key properties of such electromagnetic waves. However, to do so, we will consider the simple case of a *plane wave*.

Plane waves are common throughout science and engineering, and a plane wave is defined as one that propagates in a specific direction and is uniform in any plane perpendicular to the direction of propagation. An example is a sound wave a long way from a point source, as illustrated in Figure 7.2. A sound wave will be emitted uniformly in all directions from the point source. Therefore, if we could see the high-pressure peaks in the sound wave, they would form a series of concentric spheres moving away from the point source, with each sphere separated by a wavelength from its neighbours. A long way from the source, the radius of curvature of the sphere becomes so large that the high-pressure peaks look like flat sheets which are perpendicular to the direction of propagation (rather as the surface of the earth looks flat to us, because the radius of curvature is so much larger than the distance we can see). Therefore, for the sound wave example, at any moment in time we would measure the same pressure at any point in any single plane perpendicular to the direction of propagation – we have a plane wave. In reality, as with the example of the surface of the earth, we are assuming that the plane is small in dimensions relative to the radius of curvature of the wave, as is the case in Figure 7.2 where only a small area of the wave has been shown. Other examples of plane waves in electromagnetism include

Figure 7.2 Illustration of sound waves with a wavelength λ propagating from a point source. Close to the source, the high-pressure wave fronts appear spherical, but at a long distance from the source, they look like flat, plane waves, as long as the area we are measuring over is small compared with the radius of curvature (i.e. the distance to the source).

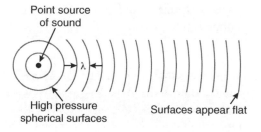

the light from the sun or a radio wave a long distance from a transmitter for the same reason as the sound wave appeared to be a plane wave.

To simplify things further, we will assume the plane wave is *monochromatic*. In other words, it can be expressed mathematically by a simple travelling sinusoidal wave with a well-defined frequency and wavelength. As any complex wave in three dimensions can be expressed as the sum of an infinite series of plane waves using Fourier analysis, the insights which we gain allow us to understand more complex scenarios as well.

Let us consider a monochromatic wave that is propagating in a dielectric medium. For convenience, we shall define the direction of propagation as being in the positive z-direction using Cartesian coordinates.

The first question that we might want to ask is whether the electric field points in any particular direction relative to the direction of propagation. A valid solution to the wave equation for the electric field (Eq. (7.7)) is

$$\mathbf{E} = (E_{0x}\mathbf{i} + E_{0y}\mathbf{j} + E_{0z}\mathbf{k}) \exp\{j(\omega t - \beta z)\} \tag{7.25}$$

where $\mathbf{i}$, $\mathbf{j}$ and $\mathbf{k}$ are the unit vectors and E_{0x}, E_{0y} and E_{0z} are the components of the amplitude of the electric field in each of the x-, y- and z-directions, respectively. This wave is clearly propagating in the positive z-direction (see Section 6.3) but is otherwise uniform in any x–y plane. This equation allows the electric field vector to point in any direction. However, we know from the Maxwell equations (Eq. (7.1a)) that in a dielectric medium $\nabla . \mathbf{D} = 0$ as there is no net charge, and therefore as $\mathbf{D} = \varepsilon_0 \varepsilon_r \mathbf{E}$, it follows that $\nabla . \mathbf{E} = 0$. In Cartesian coordinates, this can be expressed as

$$\nabla.\mathbf{E} = \frac{\partial E_x}{\partial x} + \frac{\partial E_y}{\partial y} + \frac{\partial E_z}{\partial z} = 0 \tag{7.26}$$

As Eq. (7.25) has no dependence on either x or y,

$$\frac{\partial E_x}{\partial x} = \frac{\partial E_y}{\partial y} = 0 \tag{7.27}$$

and therefore Eq. (7.26) becomes

$$\frac{\partial E_z}{\partial z} = 0 \tag{7.28}$$

Substituting in the equation for the monochromatic plane wave (Eq. (7.25)) gives

$$-j\beta E_{0z} \exp\{j(\omega t - \beta z)\} = 0 \tag{7.29}$$

which has only two possible solutions, given that the exponential terms cannot be zero for all values of z. One possible solution is that $\beta = 0$, but this implies an infinite wavelength as $\beta = 2\pi/\lambda$; this is not a wave and so cannot be valid. Therefore, the only other possible solution is that

$$E_{0z} = 0 \tag{7.30}$$

which means that no component of the electric field vector points in the direction of propagation of the electromagnetic wave. This is another key result, as consequently the electric field vector in an electromagnetic wave always points perpendicularly to the direction of propagation.

It can also be shown that the magnetic field points perpendicularly to the direction of propagation using a similar argument. In this case, the solution to the wave equation (Eq. (7.13)) is

$$\mathbf{H} = (H_{0x}\mathbf{i} + H_{0y}\mathbf{j} + H_{0z}\mathbf{k})\exp\{j(\omega t - \beta z)\} \tag{7.31}$$

From the Maxwell equations (Eq. (7.1b)), $\nabla . \mathbf{B} = 0$, and therefore in a dielectric medium $\nabla . \mathbf{H} = 0$ also. As Eq. (7.31) has no dependence on either x or y,

$$\frac{\partial H_z}{\partial z} = 0 \tag{7.32}$$

from Eq. (7.26), and therefore by substituting Eq. (7.31) into Eq. (7.32) it follows that

$$H_{0z} = 0 \tag{7.33}$$

We have shown that both the electric and magnetic fields point perpendicularly to the direction of propagation of an electromagnetic wave, but the Maxwell equations (7.1c) and (7.1d) also suggest that the electric and magnetic fields are themselves related to each other. To understand this relationship, let us consider the special case of a monochromatic plane electromagnetic wave propagating in the positive z-direction where the electric field only has a component in the x-direction. Such a wave is called a *linearly polarized* wave, and it can be expressed as

$$\mathbf{E} = E_{0x}\mathbf{i}\exp\{j(\omega t - \beta z)\} \tag{7.34}$$

Having defined the electric field component of the wave, we can now find the magnetic field component by substitution into Eq. (7.1c). For this, we shall use the fact that the curl of any vector quantity $\mathbf{Y}$ can be expressed in Cartesian coordinates as

$$\nabla \times \mathbf{Y} = \begin{vmatrix} \mathbf{i} & \mathbf{j} & \mathbf{k} \\ \partial/\partial x & \partial/\partial y & \partial/\partial z \\ Y_x & Y_y & Y_z \end{vmatrix} \tag{7.35}$$

where Y_x, Y_y and Y_z are the components of $\mathbf{Y}$ in each of the x-, y- and z-directions, respectively. Therefore, this substitution yields

$$\nabla \times \mathbf{E} = \begin{vmatrix} \mathbf{i} & \mathbf{j} & \mathbf{k} \\ \partial/\partial x & \partial/\partial y & \partial/\partial z \\ E_{0x}\exp\{j(\omega t - \beta z)\}_x & 0 & 0 \end{vmatrix} = -\frac{\partial \mathbf{B}}{\partial t} \tag{7.36}$$

Evaluation of the determinant gives

$$\mathbf{j}\frac{\partial}{\partial z}[E_{0x}\exp\{j(\omega t - \beta z)\}] - \mathbf{k}\frac{\partial}{\partial y}[E_{0x}\exp\{j(\omega t - \beta z)\}] = -\frac{\partial \mathbf{B}}{\partial t} \tag{7.37}$$

and as there is no y-dependence in the second term on the left-hand side of this expression, this reduces to

$$\mathbf{j}j\beta E_{0x}\exp\{j(\omega t - \beta z)\} = \frac{\partial \mathbf{B}}{\partial t} \tag{7.38}$$

By integrating both sides of this equation with respect to time, we get

$$\mathbf{B} = \mathbf{j}\frac{\beta}{\omega}E_{0x}\exp\{j(\omega t - \beta z)\} \tag{7.39}$$

Finally, knowing that $\mathbf{B} = \mu_0\mu_r\mathbf{H}$ in a dielectric medium, we can rewrite this as

$$\mathbf{H} = H_{0y}\mathbf{j}\exp\{j(\omega t - \beta z)\} \tag{7.40}$$

where

$$H_{0y} = \frac{\beta}{\omega}\frac{E_{0x}}{\mu_0\mu_r} \tag{7.41}$$

It is immediately clear from this equation that the magnetic field must be pointing in the y-direction. Therefore, not only do the magnetic and electric fields both point in a direction perpendicular to the direction of propagation of the wave, but they also point perpendicularly to each other. This gives us the picture of the linearly polarized, monochromatic plane electromagnetic wave shown in Figure 7.3a, with the electric field and magnetic fields varying sinusoidally as a function of the z-direction of propagation, but themselves pointing in the x- and y-directions, respectively. Figure 7.3b is another representation of the same wave, but this time highlighting the planes where the electric and magnetic fields are a maximum.

A simple way of remembering the relative orientation of the fields and the direction of propagation in an electromagnetic wave is to use the *right-hand rule*. If you point the index finger of your right hand in the direction of the electric field vector and middle finger in the

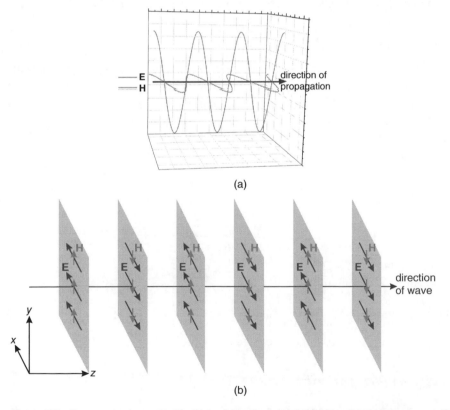

(a)

(b)

Figure 7.3 Representations of a linearly polarized, monochromatic, plane electromagnetic wave propagating in the z-direction (a) in sinusoidal form and (b) showing the wave fronts where the electric and magnetic fields are a maximum, with the directions of the fields indicated.

direction of the magnetic field, then your thumb will point in the direction of propagation if you hold it perpendicularly to the other two fingers.

7.6 Intrinsic Impedance

We have seen in Section 7.4 that there is a clear relationship between the expression for the electric and magnetic field components in a linearly polarized, monochromatic, plane electromagnetic wave, and we have considered how the direction of the electric field, magnetic field and direction of propagation of the wave relate to each other. However, it is also apparent from Eq. (7.41) that the magnitudes of the electric and magnetic fields are also quantitatively related to each other. In particular, the quantity ω/β is just the wave velocity c (Eq. (6.23)) which, from Eq. (7.9), we have already calculated for an electromagnetic wave to be $1/\sqrt{\varepsilon_0\varepsilon_r\mu_0\mu_r}$. Therefore, substituting this into Eq. (7.41) gives

$$H_{0y} = E_{0x}\sqrt{\frac{\varepsilon_0\varepsilon_r}{\mu_0\mu_r}} \tag{7.42}$$

The relative magnitude of the electric and magnetic fields is only dependent on the properties of the dielectric medium through which the wave is propagating: namely the permittivity (how readily the material can be polarized) and permeability (how readily the material can be magnetized).

Based on this, we define a new quantity called the *intrinsic impedance* of a medium as being the ratio of the magnitude of the electric and magnetic fields. It is given the symbol of the Greek letter η, and so from Eq. (7.42),

$$\eta = \frac{|\mathbf{E}|}{|\mathbf{H}|} = \sqrt{\frac{\mu_0\mu_r}{\varepsilon_0\varepsilon_r}} \tag{7.43}$$

If we remember that electric field has units of $V\,m^{-1}$ and magnetic field has units of $A\,m^{-1}$, then characteristic impedance has units of $V\,A^{-1}$, which is the same as the unit of ohms (Ω). Therefore, this is the same dimensionally as electrical impedance.

By comparing Eq. (7.43) with the equation for the characteristic impedance of an ideal transmission line Z_0 (Eq. (6.56)) it is apparent that this is related to the intrinsic impedance of the dielectric material between the conductors of the transmission line by

$$Z_0 = g\eta \tag{7.44}$$

where g is a geometry factor that is specific to the transmission line. This should be an expected result. The voltage between the conductors at any point on a transmission line is calculated by integrating the electric field along a line between the conductors, while the current is calculated by integrating the magnetic field along another line passing between the conductors (see Section 6.2). Therefore, the characteristic impedance of the transmission line Z_0 must simply be a function of the geometry of the line and the intrinsic impedance η of the dielectric.

As we shall see in most of the subsequent chapters, the intrinsic impedance of a dielectric medium is an enormously valuable quantity. It allows us to quickly determine the relative magnitudes of the electric and magnetic fields. Also, just as we can use characteristic impedance to determine how waves propagate from one transmission line to another (see Chapter 6), we can use intrinsic impedance to determine how waves behave as they propagate from one medium to another. Examples include when light passes from a semiconductor laser to an optical fibre or as light passes into the eye from air. A useful quantity to remember is the intrinsic impedance of vacuum, which is 377 Ω.

7.7   Power and the Poynting Vector

In this chapter, we have used the fact that there are no free charges in a dielectric medium to adapt the Maxwell equations and combine them to prove that electromagnetic waves can propagate through free space. These waves consist of time-varying electric and magnetic fields which can be expressed by travelling wave equations of the form of Eqs. (7.34) and (7.40) for the special case of linearly polarized, monochromatic plane electromagnetic waves. Thus far, we have only considered the properties of the electric and magnetic fields themselves, but we know that we can transmit energy using electromagnetic waves as it requires work to be done to generate an electromagnetic wave and we can do work with the energy received when we absorb an electromagnetic wave. We shall consider the generation and absorption of electromagnetic waves in a little more detail when we look at antennas in Chapter 8. However, the consequence of this is that an electromagnetic wave must itself have an energy density (or energy per unit volume of the space that it is occupying). Furthermore, as this energy is being transmitted from one place to another, if we image some area perpendicular to the direction of propagation of the wave, then there must be some flow of energy through this area, and so it must be possible to define a power per unit area of the wave. This was considered by John Henry Poynting in the late nineteenth century (Poynting 1884), and from his work we define a new quantity called the *Poynting vector*, which is given by

$$\mathbf{N} = \mathbf{E} \times \mathbf{H} \tag{7.45}$$

If the electric field and magnetic field in an electromagnetic wave are pointing in the *x*- and *y*-directions respectively, then the cross product of the two will result in a vector pointing in the *z*-direction, which is the direction of propagation of the wave; hence the Poynting vector points in this direction.

The magnitude of the Poynting vector is also significant, being the instantaneous power per unit area in the electromagnetic wave. To understand this, let us consider a small cubic volume v of space within the path of a plane electromagnetic wave travelling in the positive *z*-direction. In this scenario, there must be a flux of power into this volume through the face pointing in the negative *z*-direction and out through the opposite face, but there is no flux of power through the other faces. If there is no power dissipation or generation inside the small volume, then the total flux of power through the surface surrounding the small volume must be zero. From our consideration of the energy in electric and magnetic fields in Section 5.6, the volume must contain some energy U at any moment in time. Let us calculate the integral of the Poynting vector passing through the closed surface that bounds the small volume by starting with Eq. (7.45), so that

$$\oint_S \mathbf{N}.\mathbf{dA} = \oint_S (\mathbf{E} \times \mathbf{H}).\mathbf{dA} \tag{7.46}$$

We have already seen that the Gauss theorem for divergence allows a surface integral to be converted into a volume integral such that for any arbitrary vector quantity **Y**,

$$\int_V (\nabla . \mathbf{Y}) dV = \oint_S \mathbf{Y} . \mathbf{dA} \tag{5.3}$$

Therefore, Eq. (7.46) can be rewritten as

$$\oint_S \mathbf{N} . \mathbf{dA} = \int_V [\nabla . (\mathbf{E} \times \mathbf{H})] dV \tag{7.47}$$

However, we can use the standard vector calculus identity

$$\nabla . (\mathbf{E} \times \mathbf{H}) = \mathbf{H} . (\nabla \times \mathbf{E}) - \mathbf{E} . (\nabla \times \mathbf{H}) \tag{7.48}$$

into which we can substitute two of the Maxwell equations in dielectric media (the Faraday law, Eq. (7.1c), and the Ampère–Maxwell law, Eq. (7.1d)) to give

$$\nabla . (\mathbf{E} \times \mathbf{H}) = -\mathbf{H} . \left(\frac{\partial \mathbf{B}}{\partial t} \right) - \mathbf{E} . \left(\frac{\partial \mathbf{D}}{\partial t} \right) \tag{7.49}$$

As we are dealing with a dielectric medium where Eqs. (2.15) and (4.9) apply,

$$\nabla . (\mathbf{E} \times \mathbf{H}) = -\frac{\mathbf{B}}{\mu_0 \mu_r} . \left(\frac{\partial \mathbf{B}}{\partial t} \right) - \frac{\mathbf{D}}{\varepsilon_0 \varepsilon_r} . \left(\frac{\partial \mathbf{D}}{\partial t} \right) \tag{7.50}$$

However, from the chain rule, we know that for any vector quantity **Y**,

$$\frac{\partial (\mathbf{Y} . \mathbf{Y})}{\partial t} = 2\mathbf{Y} . \frac{\partial (\mathbf{Y})}{\partial t}$$

Therefore, with this transformation applied to Eq. (7.50) and substitution into Eq. (7.47), we are left with

$$\oint_S \mathbf{N} . \mathbf{dA} = -\frac{1}{2} \int_V \left[\frac{1}{\mu_0 \mu_r} . \left(\frac{\partial (\mathbf{B} . \mathbf{B})}{\partial t} \right) + \frac{1}{\varepsilon_0 \varepsilon_r} . \left(\frac{\partial (\mathbf{D} . \mathbf{D})}{\partial t} \right) \right] dV \tag{7.51}$$

which in terms of electric and magnetic fields becomes

$$\oint_S \mathbf{N} . \mathbf{dA} = -\int_V \left[\frac{\partial}{\partial t} \left(\frac{\mathbf{B} . \mathbf{H}}{2} + \frac{\mathbf{D} . \mathbf{E}}{2} \right) \right] dV \tag{7.52}$$

From Eq. (2.27) we know that **D** . **E**/2 is the energy stored per unit volume in a dielectric due to an electric field and from Eq. (4.40) that **B** . **H**/2 is the energy stored per unit volume due to a magnetic field, as we saw in Section 5.6. Therefore, the right-hand side of Eq. (7.52) is the negative of the rate of change of total energy stored, U, within the volume bounded by the surface integral on the left-hand side of the equation:

$$\oint_S \mathbf{N} . \mathbf{dA} = -\int_V \frac{\partial U}{\partial t} dV \tag{7.53}$$

In other words, it is the power leaving the volume that is passing out through the surface. Therefore, the Poynting vector **N** must be the power per unit area in the electromagnetic wave.

To understand the quantitative significance of the Poynting vector further, let us consider the monochromatic, linearly polarized, plane electromagnetic wave from Section 7.4 whose electric and magnetic fields are expressed by Eqs. (7.34) and (7.40), respectively. It is, of

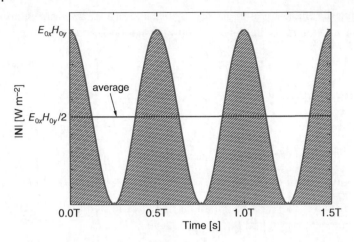

Figure 7.4 The instantaneous power per unit area given by |**N**| for the monochromatic, linearly polarized plane electromagnetic wave, plotted as a function of time at $z = 0$. The power oscillates between a maximum of $E_{0x}H_{0y}$ and zero, with the average power being half of the peak power.

course, only the real part of these equations which represents the actual fields, and so we can evaluate the Poynting vector to be

$$\mathbf{N} = \mathbf{E} \times \mathbf{H} = \begin{vmatrix} \mathbf{i} & \mathbf{j} & \mathbf{k} \\ E_{0x}\cos(\omega t - \beta z) & 0 & 0 \\ 0 & H_{0y}\cos(\omega t - \beta z) & 0 \end{vmatrix}$$

$$\mathbf{N} = \mathbf{k}E_{0x}H_{0y}\cos^2(\omega t - \beta z) \tag{7.54}$$

This confirms that the wave is propagating in the positive z-direction and the power per unit area varies with both time and position. For example, |**N**| is plotted as a function of time at $z = 0$ in Figure 7.4. This shows that the power is oscillating with time between a maximum of $E_{0x}H_{0y}$ and zero. As the integral of $\cos^2(\omega t - \beta z)$ over a complete cycle is 1/2, the average power per unit area $\overline{|\mathbf{N}|}$ must be only half the peak power and in this case is given by

$$\overline{|\mathbf{N}|} = \frac{E_{0x}H_{0y}}{2} \tag{7.55}$$

although more generally it is

$$\overline{|\mathbf{N}|} = \frac{|\mathbf{E}||\mathbf{H}|}{2} \tag{7.56}$$

If we are using a complex number notation, to ensure that we only take the real part of the fields we must use

$$\overline{|\mathbf{N}|} = \frac{\mathbf{E} \times \mathbf{H}^*}{2} \tag{7.57}$$

It is worth noting the similarity with a.c. electrical power, where the average power is half of the peak power and therefore root mean square (rms) values of current and voltage are frequently used in order to make calculations of power straightforward. Likewise, we can define rms values $E_{rms} = |\mathbf{E}|/\sqrt{2}$ and $H_{rms} = |\mathbf{H}|/\sqrt{2}$ of the electric and magnetic fields respectively such that

$$\overline{|\mathbf{N}|} = E_{rms}H_{rms} \tag{7.58}$$

In practice, it is the average magnitude of the Poynting vector that is of most practical use in calculations where typically we wish to know the total power that is being either emitted or received at different points in space, as we shall see when we consider antennas in the next chapter.

Reference

Poynting, J.H. (1884). XV. On the transfer of energy in the electromagnetic field. *Philosophical Transactions of the Royal Society of London* 175: 343–361.

8

Antennas

8.1 Introduction

In calculating the power in an electromagnetic wave in Section 7.6, we showed that the Poynting vector **N** was the instantaneous power per unit area by finding that the integral of the Poynting vector leaving a closed surface was equal to the negative of the rate of change of energy stored in the enclosed volume of space (Eq. (7.53)). We did not allow for the presence of an energy source inside the volume in this derivation that might be outputting power P_o in the form of an electromagnetic wave or for a mechanism for dissipating power P_i from an electromagnetic wave. If we include these terms, then Eq. (7.53) becomes

$$\oint_S \mathbf{N} \cdot \mathbf{dA} = -\int_V \frac{\partial U}{\partial t} dV + P_o - P_i \tag{8.1}$$

In Chapter 10, we shall consider the interaction of electromagnetic waves with conducting media and shall see that in this scenario the free charges that are present dissipate power in electromagnetic waves. However, in this chapter we shall look specifically at how we engineer *antennas* which can convert an electrical signal into an electromagnetic wave and vice versa.

We are familiar with the landscape around us being dotted with large antennas for transmitting television and radio signals, and with receiving antennas on the roofs of houses and cars. We also see mobile phone masts which have both transmitting and receiving antennas, but perhaps we often forget that the mobile telephone in our pockets also contains both transmitting and receiving antennas, not only to communicate with the mobile phone network, but also for Wi-Fi, Bluetooth® and other communications standards.

A diversity of antenna designs are therefore required, depending on the frequency of the electromagnetic wave, the power requirements, directionality and whether it is a transmitter or receiver. In this chapter, we shall look at the physics underlying the analysis of antennas and apply this to the most basic of antenna designs – the dipole antenna. We shall then draw up some basic figures of merit for comparing antennas of different designs quantitatively.

Electromagnetism for Engineers, First Edition. Andrew J. Flewitt.
© 2023 John Wiley & Sons Ltd. Published 2023 by John Wiley & Sons Ltd.
Companion website: www.wiley.com/go/flewitt/electromagnetism

8.2 Retarded Potentials

In this chapter, we will want to know what electric and magnetic field results at an arbitrary point in space defined by some vector **r** from a transmitting antenna which is itself at a position $\mathbf{r}_0$ as shown in Figure 8.1 (where both vectors are with respect to some defined origin in space). It is clear that the Maxwell equations will need to form the basis of such a calculation, but so far we have almost exclusively used these in terms of electric and magnetic fields, as in Table 5.1. For this problem, it turns out to be more convenient to use a potential-based approach.

We have already discussed the electric potential V in detail in Section 1.3, where the difference in potential V_{ab} between two points, a and b, in space was defined as the change in energy per unit charge between the two and can be calculated from the integral of the electric field between the two points:

$$V_{ab} = -\int_b^a E\,dx \tag{1.8}$$

Conversely, the electric field is the negative of the gradient in the electric potential:

$$\mathbf{E} = -\nabla V \tag{1.7}$$

Furthermore, there is no fundamental definition of zero potential – we are free to define this for any particular problem, although for many practical situations we consider the earth to be at zero potential.

We can rewrite Eq. (1.7) in terms of the electric flux density if we know the relative permittivity of the medium through which the field is passing using Eq. (2.15):

$$\mathbf{D} = -\varepsilon_0 \varepsilon_r \nabla V \tag{8.2}$$

Substitution of this into the Gauss law of electric fields in differential form (Eq. (5.7)) gives

$$\nabla^2 V = \frac{-\rho}{\varepsilon_0 \varepsilon_r} \tag{8.3}$$

This is commonly called the *Poisson equation*, and it allows the electric potential to be calculated from a charge distribution. From this basis, we have already shown that the electric potential around a point charge is given by Eq. (1.10), and therefore we can calculate the potential at any point **r** as a result of some complex, arbitrary charge distribution on an

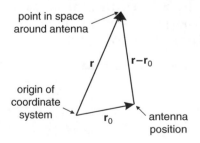

Figure 8.1 The coordinate system under consideration consists of an antenna at some position $\mathbf{r}_0$ relative to the origin where there is some point charge or point current. We wish to know the potential at some other arbitrary point **r**.

antenna $\rho(\mathbf{r}_0)$ by considering this to be a number of point charges which can simply be integrated, so that

$$V(\mathbf{r}) = \frac{1}{4\pi\varepsilon_0\varepsilon_r} \int \frac{\rho(\mathbf{r}_0)}{|\mathbf{r} - \mathbf{r}_0|} dv \tag{8.4}$$

where the integral is over the volume occupied by the charge distribution.

We have assumed that this charge distribution is static, but if it is time-varying (e.g. at a transmitting antenna) then it takes some finite time for the effect of the changing charge distribution at $\mathbf{r}_0$ to propagate to point $\mathbf{r}$ (see Figure 8.1). This effective time delay will be determined by the speed of the electromagnetic wave c in the medium and will be equal to $|\mathbf{r} - \mathbf{r}_0|/c$. Therefore, Eq. (8.4) must be rewritten to allow for this, so that

$$V(\mathbf{r}, t) = \frac{1}{4\pi\varepsilon_0\varepsilon_r} \int \frac{\rho(\mathbf{r}_0, t - |\mathbf{r} - \mathbf{r}_0|/c)}{|\mathbf{r} - \mathbf{r}_0|} dv \tag{8.5}$$

This is known as a *retarded potential*, and it is effectively saying that the potential at any point away from an antenna is dependent on the charge distribution at the antenna at the earlier moment in time when the wave was emitted.

While Eq. (8.5) allows us to determine the effect of the electric component of an electromagnetic wave, it does not contain information about the magnetic component. We need an expression similar to Eq. (8.5), but for this we require a magnetic equivalent to the electric potential, and we call this the *magnetic vector potential*. Just as the electric potential is able to uniquely define the electric field (although the electric field does not uniquely define the electric potential), so we require the same property of the magnetic vector potential. We know from the Maxwell equations that $\nabla \cdot \mathbf{B} = 0$ (Eq. (5.9)), while from vector calculus the divergence of the curl of any vector quantity is zero. Therefore, for any magnetic flux density, we must be able to define another vector quantity $\mathbf{A}$ where

$$\mathbf{B} = \nabla \times \mathbf{A} \tag{8.6}$$

so that

$$\nabla \cdot \mathbf{B} = \nabla \cdot (\nabla \times \mathbf{A}) = 0 \tag{8.7}$$

thereby satisfying the Maxwell equation (Eq. (5.9)). It is this quantity $\mathbf{A}$ that is the magnetic vector potential. Unlike the electric potential, it does not have a simple physical meaning, but it is very helpful for the purposes of calculations and particularly those involving time-varying currents, as is the case for antennas. Like the electric potential, although $\mathbf{A}$ uniquely defines $\mathbf{B}$, the reverse is not true. However, to keep calculations as simple as possible, it is convenient to make the additional constraint that

$$\nabla \cdot \mathbf{A} = 0 \tag{8.8}$$

which does then give a unique solution to $\mathbf{A}$ for any $\mathbf{B}$.

We can now derive a magnetic equivalent to the Poisson equation (Eq. (8.3)) by rewriting the Ampère–Maxwell law

$$\nabla \times \mathbf{H} = \mathbf{J} + \frac{\partial \mathbf{D}}{\partial t} \tag{5.32}$$

in terms of the magnetic vector potential. However, just as the Poisson equation was derived for static (not time-varying) fields, so we do the same here, such that $\partial \mathbf{D}/\partial t = 0$. Therefore, if we know the permeability of the medium through which the field is passing then we can substitute Eq. (8.6) into Eq. (5.32) to give

$$\nabla \times \nabla \times \mathbf{A} = \mu_0 \mu_r \mathbf{J} \tag{8.9}$$

However, from the standard vector calculus identity, we know that

$$\nabla \times \nabla \times \mathbf{A} = \nabla(\nabla \cdot \mathbf{A}) - \nabla^2 \mathbf{A} = \mu_0 \mu_r \mathbf{J} \tag{8.10}$$

Finally, substituting in our additional uniqueness constraint (Eq. (8.8)),

$$\nabla^2 \mathbf{A} = -\mu_0 \mu_r \mathbf{J} \tag{8.11}$$

which is clearly equivalent to the Poisson equation and allows the magnetic vector potential to be determined for any current distribution.

Using the same coordinate system as in Figure 8.1, we can use Eq. (8.11) to calculate the magnetic vector potential at any position $\mathbf{r}$ around a 'point current' $\mathbf{J}$ due to a moving charge at position $\mathbf{r}_0$ to be

$$\mathbf{A}(\mathbf{r}) = \frac{\mu_0 \mu_r \mathbf{J}}{4\pi |\mathbf{r} - \mathbf{r}_0|} \tag{8.12}$$

which has a very similar form to the electric equivalent (Eq. (1.10)). Therefore, we can calculate the magnetic potential at any point $\mathbf{r}$ as a result of some complex, arbitrary current distribution on an antenna $\mathbf{J}(\mathbf{r}_0)$ by considering this to be a number of point currents which can simply be integrated, so that

$$\mathbf{A}(\mathbf{r}) = \frac{\mu_0 \mu_r}{4\pi} \int \frac{\mathbf{J}(\mathbf{r}_0)}{|\mathbf{r} - \mathbf{r}_0|} dv \tag{8.13}$$

where the integral is over the volume occupied by the current distribution, to give an equation that is analogous to Eq. (8.4). Once again, we have derived this equation for a situation that is static (i.e. the current distribution is not changing with time). For the case of an antenna where the current is time-dependent, we again need to allow for the finite time for the effect of the changing current to propagate through space at the velocity of the electromagnetic wave in the medium c. Therefore, we need to rewrite Eq. (8.13) to allow for this,

$$\mathbf{A}(\mathbf{r}, t) = \frac{\mu_0 \mu_r}{4\pi} \int \frac{\mathbf{J}(\mathbf{r}_0, t - |\mathbf{r} - \mathbf{r}_0|/c)}{|\mathbf{r} - \mathbf{r}_0|} dv \tag{8.14}$$

which, like Eq. (8.5), is also known as a retarded potential. We shall now see how these retarded potentials can be used to calculate the transmission of electromagnetic waves by antennas.

8.3   Short Dipole Antenna

The simplest form of transmitting antenna is the *dipole antenna* which consists of a short length l of a conducting wire to which a signal for broadcast is applied as a time-varying current. In practice, the exact form of this current will depend on the specific application (e.g. AM radio broadcasting in which a high-frequency sine wave is modulated in amplitude to encode an audio signal). However, in all cases we can break down a complex waveform into an infinite series of sine waves using Fourier analysis. Therefore, we can study the simple case of a sinusoidal current into the antenna as shown in Figure 8.2, of the form

$$I - I_0 \cos(\omega t) \tag{8.15}$$

where I_0 is the amplitude of the current and ω is the angular frequency, knowing that the real situation would be a superposition of
many such simple signals. Implicit in this expression for the current
is that the antenna is sufficiently short for the current to be uniform
at any moment in time along the length – in other words, the antenna
is short compared with the wavelength. For a long dipole antenna,
the current would need to have position-dependent terms as well.

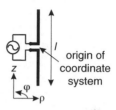

Figure 8.2 The
dipole antenna of
length l at the origin
of a cylindrical polar
coordinate system.

We can use the retarded potential (Eq. (8.14)) to calculate the magnetic vector potential around the antenna. If we define the origin of
a cylindrical polar coordinate space to be at the antenna (Figure 8.2)
so that $\mathbf{r}_0 = 0$ at the centre of the antenna and we only consider a
far-field scenario where $|\mathbf{r}| \gg l$ then Eq. (8.14) simplifies to

$$\mathbf{A}(\mathbf{r}, t) = \frac{\mu_0 \mu_r}{4\pi |\mathbf{r}|} \int \mathbf{J}(\mathbf{r}_0, t - |\mathbf{r}|/c) dv \tag{8.16}$$

Here the integral is over the line of current along the length of the wire. Also, if we define
the coordinate system with the z-axis along the line of the current, so that $\mathbf{J}$ only has a
z-component, then it is clear from Eq. (8.16) that $\mathbf{A}$ will also only have a z-component,

$$A_z(\mathbf{r}, t) = \frac{\mu_0 \mu_r}{4\pi |\mathbf{r}|} \int_{-l/2}^{l/2} I(t - |\mathbf{r}|/c) dz \tag{8.17}$$

As I is the same at all points along the wire at any moment in time, this becomes

$$A_z(\mathbf{r}, t) = \frac{\mu_0 \mu_r l}{4\pi |\mathbf{r}|} I_0 \cos(\omega[t - |\mathbf{r}|/c]) \tag{8.18}$$

For simplicity, we can express this in complex notation as

$$A_z(\mathbf{r}, t) = \frac{\mu_0 \mu_r l}{4\pi |\mathbf{r}|} I_0 \exp(j[\omega t - \beta |\mathbf{r}|]) \tag{8.19}$$

where the propagation constant $\beta = \omega/c$ (Eq. (6.23)). Finally, as $|\mathbf{r}|^2 = \rho^2 + z^2$ in cylindrical
polar coordinates, we arrive at the complete solution for the magnetic vector potential:

$$A_z(\rho, \varphi, z, t) = \frac{\mu_0 \mu_r l}{4\pi (\rho^2 + z^2)^{1/2}} I_0 \exp(j[\omega t - \beta(\rho^2 + z^2)^{1/2}]) \tag{8.20a}$$

$$A_\rho(\rho, \varphi, z, t) = 0 \tag{8.20b}$$

$$A_\varphi(\rho, \varphi, z, t) = 0 \tag{8.20c}$$

From this, we can determine the magnetic flux density using Eq. (8.6), which in cylindrical polar coordinates becomes

$$\mathbf{B} = \left(\frac{1}{\rho} \frac{\partial A_z}{\partial \varphi} - \frac{\partial A_\varphi}{\partial z} \right) \hat{\rho} + \left(\frac{\partial A_\rho}{\partial z} - \frac{\partial A_z}{\partial \rho} \right) \hat{\varphi} + \frac{1}{\rho} \left(\frac{\partial [\rho A_\varphi]}{\partial \rho} - \frac{\partial A_\rho}{\partial \varphi} \right) \hat{z} \tag{8.21}$$

where $\hat{\rho}$, $\hat{\varphi}$, and $\hat{z}$ are the unit vectors in each of the cylindrical polar axes. It is apparent from Eq. (8.20a) that A_z has no dependence on φ, so $\partial A_z / \partial \varphi = 0$. Substituting this with Eqs. (8.20b) and (8.20c) immediately simplifies Eq. (8.21) to

$$\mathbf{B} = -\frac{\partial A_z}{\partial \rho} \hat{\varphi} \tag{8.22}$$

Therefore, this can be evaluated from Eq. (8.20) to be

$$B_z(\rho, \varphi, z, t) = 0 \tag{8.23a}$$

$$B_\rho(\rho, \varphi, z, t) = 0 \tag{8.23b}$$

$$B_\varphi(\rho, \varphi, z, t) = \frac{\rho \mu_0 \mu_r I I_0 \exp(j[\omega t - \beta(\rho^2 + z^2)^{1/2}])}{4\pi(\rho^2 + z^2)} \left\{ j\beta + \frac{1}{(\rho^2 + z^2)^{1/2}} \right\} \tag{8.23b}$$

These equations collectively show that the magnetic flux density always points in the azimuthal direction and has no component in either the height or radial directions. Furthermore, the ρ term in the numerator of Eq. (8.23c) shows that the field is zero along the line of the dipole antenna itself.

In calculating the magnetic vector potential, we have already assumed that we are only interested in far-field conditions. In this case, we are many wavelengths away from the antenna, and so $(\rho^2 + z^2)^{1/2} \gg \lambda$. As $\beta = 2\pi/\lambda$, it follows that $j\beta \gg 1/(\rho^2 + z^2)^{1/2}$, and so Eq. (8.23) simplifies to

$$B_z(\rho, \varphi, z, t) = 0 \tag{8.24a}$$

$$B_\rho(\rho, \varphi, z, t) = 0 \tag{8.24b}$$

$$B_\varphi(\rho, \varphi, z, t) = \frac{j\beta \rho \mu_0 \mu_r I I_0}{4\pi(\rho^2 + z^2)} \exp(j[\omega t - \beta(\rho^2 + z^2)^{1/2}]) \tag{8.24c}$$

While working in cylindrical polar coordinates is helpful for considering the vector potential and magnetic flux density, it is more insightful to work in spherical polar coordinates when calculating the electric field, where $\hat{r}$, $\hat{\theta}$, and $\hat{\varphi}$ are the unit vectors in each of the spherical polar axes. The standard transformations between these coordinate systems (Stroud 1996) allow the magnetic flux density to be rewritten as

$$B_r(r, \theta, \varphi, t) = 0 \tag{8.25a}$$

$$B_\theta(r, \theta, \varphi, t) = 0 \tag{8.25b}$$

$$B_\varphi(r, \theta, \varphi, t) = \frac{j\beta \mu_0 \mu_r I I_0 \sin\theta}{4\pi r} \exp(j[\omega t - \beta r]) \tag{8.25c}$$

We can now derive the electric field from the magnetic flux density using the Ampère–Maxwell equation in dielectrics (Eq. (7.1d))

$$\nabla \times \mathbf{B} = \mu_0 \mu_r \varepsilon_0 \varepsilon_r \frac{\partial \mathbf{E}}{\partial t} \tag{8.26}$$

In cylindrical polar coordinates,

$$\nabla \times \mathbf{B} = \frac{1}{r\sin\theta} \left(\frac{\partial[B_\varphi \sin\theta]}{\partial\theta} - \frac{\partial B_\theta}{\partial\varphi} \right) \hat{r} + \frac{1}{r} \left(\frac{1}{\sin\theta} \frac{\partial B_r}{\partial\varphi} - \frac{\partial[rB_\varphi]}{\partial r} \right) \hat{\theta}$$
$$+ \frac{1}{r} \left(\frac{\partial[rB_\theta]}{\partial r} - \frac{\partial B_r}{\partial\theta} \right) \hat{\varphi} \tag{8.27}$$

but again we have a significant simplification when we substitute in Eqs. (8.25a) and (8.25b):

$$\nabla \times \mathbf{B} = \frac{1}{r \sin \theta} \frac{\partial [B_\varphi \sin \theta]}{\partial \theta} \, \hat{\mathbf{r}} - \frac{1}{r} \frac{\partial [r B_\varphi]}{\partial r} \hat{\boldsymbol{\theta}} \tag{8.28}$$

$$= \frac{j\beta \mu_0 \mu_r l I_0 \cos \theta}{2\pi r^2} \exp(j[\omega t - \beta r]) \, \hat{\mathbf{r}} - \frac{\beta^2 \mu_0 \mu_r l I_0 \sin \theta}{4\pi r} \exp(j[\omega t - \beta r]) \hat{\boldsymbol{\theta}} \tag{8.29}$$

In the far field, $1/r \gg 1/r^2$, and so this simplifies to

$$\nabla \times \mathbf{B} = -\frac{\beta^2 \mu_0 \mu_r l I_0 \sin \theta}{4\pi r} \exp(j[\omega t - \beta r]) \hat{\boldsymbol{\theta}} \tag{8.30}$$

Therefore, substitution into Eq. (8.26) and integration with respect to time, which effectively means dividing by $j\omega$, allows the electric field to be calculated:

$$E_r(r, \theta, \varphi, t) = 0 \tag{8.31a}$$

$$E_\theta(r, \theta, \varphi, t) = \frac{j\beta l I_0 \sin \theta}{4\pi c \varepsilon_0 \varepsilon_r r} \exp(j[\omega t - \beta r]) \tag{8.31b}$$

$$E_\varphi(r, \theta, \varphi, t) = 0 \tag{8.31c}$$

where the substitution $c = \omega/\beta$ for waves in dielectric media (Eq. (6.23)) has been made.

These equations show that there is no component of the electric field pointing in the same direction as the magnetic field – the two are orthogonal to each other. This is the condition for propagation of an electromagnetic wave in free space, as we saw in Section 7.5. Furthermore, as neither field has a component in the radial direction, it can be concluded that the direction of propagation of the wave is radially away from the transmitting antenna, as would be intuitively expected. We should be able to calculate the average power in this radial wave from the Poynting vector using Eq. (7.57):

$$\overline{|\mathbf{N}|} = \frac{\mathbf{E} \times \mathbf{H}^*}{2} \tag{7.57}$$

In spherical polar coordinates, this becomes

$$\overline{|\mathbf{N}|} = \frac{E_r B_\varphi^*}{2\mu_0 \mu_r} \hat{\mathbf{r}} \tag{8.32}$$

$$= \frac{\beta^2 l^2 I_0^2 \sin^2 \theta}{32\pi^2 c \varepsilon_0 \varepsilon_r r^2} \hat{\mathbf{r}} \tag{8.33}$$

We can learn several things about the electromagnetic wave that is transmitted from the dipole antenna from this expression. Firstly, there is no dependence of the Poynting vector on φ, and therefore the transmitted wave is symmetric around a line through the antenna. Secondly, the power per unit area decreases as r^{-2}, which is exactly as we would expect if the medium through which the wave is travelling is not dissipating any power. Thirdly, the $\sin^2 \theta$ term shows that no power is transmitted along the line of the antenna, but instead it is preferentially transmitted in the plane perpendicular to the dipole antenna that passes through its centre.

We therefore end up with the complete picture of the dipole antenna illustrated in Figure 8.3. The magnetic field circulates in latitudinal loops that are in planes perpendicular to the antenna, while the electric field circulates in longitudinal loops in planes

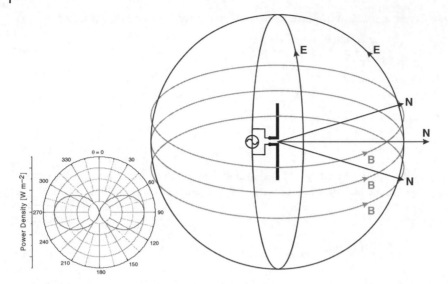

Figure 8.3 Schematic of the direction of the electric and magnetic fields and the Poynting vector around a dipole antenna that is driven by an a.c. voltage source. The inset shows how the power density varies with θ around the antenna. The power density is greatest in the plane perpendicular to the antenna passing through its centre.

aligned with the antenna. Both fields decay with distance away from the antenna with a r^{-1} relationship. The Poynting vector therefore always points away from the antenna. It decays as r^{-2} with distance and is greatest in the plane perpendicular to the antenna passing through its centre.

8.4 Antenna Figures of Merit

The short dipole antenna that we have studied in the previous section is the simplest form of transmitting antenna. However, we have seen that calculation of the transmitted power is not trivial, even in this case, as the result is highly anisotropic. There are many different types of both transmitting and receiving antenna designs; therefore, it is helpful to be able to extract some simple figures of merit which then allow the appropriate antenna to be selected for different applications.

The most basic of these figures of merit is the *antenna gain*. This is the ratio of the peak power per unit area, P_{max}, in the transmitted electromagnetic wave to the power per unit area that would result if the antenna was transmitting isotropically P_{iso}, and hence is given by

$$G = \frac{P_{max}}{P_{iso}} \tag{8.34}$$

as illustrated in Figure 8.4 for the dipole antenna. For this case, we can see from Eq. (8.33) that the Poynting vector is maximized when $\sin\theta = 1$, and so

$$P_{max} = \overline{|\mathbf{N}|}(\theta = 90°) = \frac{\beta^2 l^2 I_0^2}{32\pi^2 c\varepsilon_0\varepsilon_r r^2} \tag{8.35}$$

The isotropic power per unit area can be found by integrating $|\mathbf{N}|$ over a sphere of radius r and then dividing by the surface area of the sphere, such that

$$P_{\text{iso}} = \frac{1}{4\pi r^2} \int_0^{2\pi} \int_0^{\pi} |\mathbf{N}| r^2 \sin\theta \, d\theta \, d\varphi \tag{8.36}$$

For the case of the short dipole antenna, this becomes

$$P_{\text{iso}} = \frac{\beta^2 l^2 I_0^2}{128\pi^3 c \varepsilon_0 \varepsilon_r r^2} \int_0^{2\pi} \int_0^{\pi} \sin^3\theta \, d\theta \, d\varphi$$

$$= \frac{\beta^2 l^2 I_0^2}{48\pi^2 c \varepsilon_0 \varepsilon_r r^2} \tag{8.37}$$

Therefore, substitution of Eqs. (8.35) and (8.37) into Eq. (8.34) allows the gain for the short dipole antenna to be calculated to be $G = 1.5$. In general, a high gain implies that the antenna is transmitting a highly directional electromagnetic wave. This is often essential. For example, in satellite communications a signal only needs to be directed at a portion of the surface of the earth, which from a geostationary orbit represents a very small solid angle. By directing the transmitted wave in this way, it is possible to have a clear signal at the surface of the earth while minimizing the power required by the satellite.

It is clear that real power is being dissipated by a transmitting antenna in the form of the radiated electromagnetic wave. This power is supplied by the a.c. voltage source that is providing the input signal to the antenna, and which is shown in Figure 8.3. The antenna must appear electrically to behave like a resistance from the perspective of the voltage source. We can therefore define the *radiation resistance* R_a of the antenna as another figure of merit where

$$R_a = \frac{P_{\text{tr}}}{I_{\text{rms}}^2} \tag{8.38}$$

in which P_{tr} is the power transmitted by the antenna as an electromagnetic wave and I_{rms} is the rms current into the antenna such that this is the equivalent of the resistance that would need to be connected to the voltage source to dissipate the same real power. For the case of the short dipole antenna, the transmitted power can be calculated from the isotropic power per unit area by integrating this over a sphere at some distance r from the transmitter so that $P_{\text{tr}} = 4\pi r^2 P_{\text{iso}}$ which, from Eq. (8.37), gives

$$P_{\text{iso}} = \frac{\beta^2 l^2 I_0^2}{12\pi c \varepsilon_0 \varepsilon_r} \tag{8.39}$$

Furthermore, the rms current is related to the peak current by $I_{\text{rms}} = I_0/\sqrt{2}$, and by substituting into Eq. (8.38) we have that

$$R_a = \frac{\beta^2 l^2}{6\pi c \varepsilon_0 \varepsilon_r} \tag{8.40}$$

This expression shows that the radiation resistance is dependent on the properties of the dielectric medium through which the transmitted wave is propagating, as both the speed of the wave c and the relative permittivity ε_r appear in this equation. In fact, from Eq. (7.9) we know that $c = 1/\sqrt{\varepsilon_0 \varepsilon_r \mu_0 \mu_r}$, and therefore $1/c\varepsilon_0 \varepsilon_r = \sqrt{\mu_0 \mu_r / \varepsilon_0 \varepsilon_r}$ which itself is just the intrinsic impedance of the medium η (see Eq. (7.43)). In addition, there is a wavelength dependence as $\beta = 2\pi/\lambda$, and so the radiation resistance for the short dipole antenna can be rewritten as

$$R_a = \frac{2\pi l^2}{3\lambda^2}\eta \tag{8.41}$$

Therefore, the radiation resistance is just the characteristic impedance of the medium around the antenna multiplied by a factor that is dependent on the geometry of the antenna and the wavelength of the transmitted wave. If the antenna is being fed by a signal from a transmission line, then the radiation resistance is the load resistance at the end of the line, and some form of impedance matching will be required to ensure reflections back along the line are minimized (see Section 6.6).

Thus far, we have only considered the properties of transmitting antennas. However, we also need receiving antennas to be able to extract some of the energy from the electromagnetic wave that has been transmitted by another antenna some distance away. The exact mechanism by which this extraction takes place depends on the specific design of the antenna. In the case of the short dipole antenna, the oscillating electric field sets up an a.c. potential difference along the length of the antenna, and so the receiving antenna behaves like an ideal voltage source. However, this will also cause an alternating current on the receiving antenna which is the condition for transmission. Therefore, some of the absorbed power will be retransmitted. We know that this transmitted power can be modelled as the energy loss in the radiation resistance (see Eq. (8.38)), and therefore the receiving antenna behaves like an ideal voltage source and series radiation resistance. If the antenna is then connected to a load impedance Z_l, which would usually be the input impedance of an amplifier circuit, then, from the maximum power transfer theorem, the power dissipated in the load impedance will be maximized when $R_a = |Z_l|$. Under this optimum condition, half of the total absorbed power is dissipated in the load, but half is still retransmitted.

The total absorbed power P_r at the receiving antenna can also be calculated. If we have characterized the radiation pattern of the electromagnetic wave around the transmitting antenna, then we can calculate the power per unit area $\overline{|\mathbf{N}|}$ at the receiving antenna if we know its relative position. We can therefore define an *effective area* A_r of the receiving antenna, where

$$P_r = A_r\overline{|\mathbf{N}|} \tag{8.42}$$

The effective area is therefore the area perpendicular to the direction of propagation of the electromagnetic wave from which it would be needed to extract all of the available energy in order to have a total received power of P_r.

8.5 Practical Antenna Systems

We can now put together an electrical equivalent circuit of a combination of a transmitting and receiving antenna, as shown in Figure 8.4. A signal to be transmitted is produced by an a.c. source with an rms voltage of V_{sig} and an output impedance of Z_0 so that it matches the characteristic impedance Z_0 of the transmission line to which it is connected; this ensures maximum power transfer of the signal onto the line is achieved. The other end

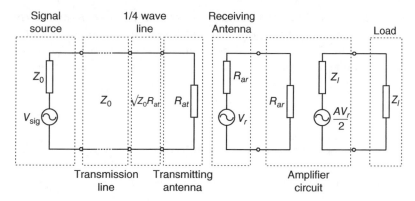

Figure 8.4 The equivalent circuit of a simple system using antennas to transmit a signal with the main functional blocks identified.

of the transmission line will need to be connected to the transmitting antenna, but as this will have an effective radiation resistance of R_{at}, where it is likely that $R_{at} \neq Z_0$. We will need to use a quarter-wave line of impedance $\sqrt{Z_0 R_{at}}$ to avoid reflections at this point (see Section 6.7). The transmitted electromagnetic wave will couple into the receiving antenna where it will induce an a.c. signal with an rms voltage of V_r. This antenna has a radiation resistance of R_{ar}. As we will immediately need to increase the signal, we will need to connect an amplifier circuit to the output of the receiving antenna, and this should have an input impedance of R_{ar} also to ensure maximum power transfer into the amplifier. If the amplifier has a gain of A, then this will produce an amplified a.c. signal of $AV_r/2$ as a result of half of the power received being retransmitted by the receiving antenna. Finally, the output impedance of the amplifier should be matched to that of the load impedance Z_l of the device in which the signal is being used, such as a radio demodulation circuit.

Appropriate selection of both transmitting and receiving antenna is clearly important in the design of a complete system. A consideration of all of the numerous designs of antennas is beyond the scope of this book. The specific combination of antennas that are actually chosen will depend on the frequency, bandwidth, power requirements, size constraints and range, to name but a few considerations. However, the *principle of reciprocity* does provide some assistance (Neiman 1943). This states that the properties of a transmitter used as a receiver are the same as its properties if used as a transmitter. We have already seen one aspect of this in the radiation resistance. However, its influence is rather wider than just this. Just as a transmitting antenna will have an anisotropic power distribution with the greatest power in a given direction, so an antenna will have the same anisotropy when used as a receiver, being most sensitive to an electromagnetic wave arriving from a specific direction. The consequence of this can be understood by considering two antennas, A and B. When A is used as a transmitter, a voltage V_A applied across its radiation resistance results in a current I_B in receiving antenna B. Similarly, when B is used as a transmitter, a voltage V_B applied across its radiation resistance results in a current I_B in receiving antenna A. From the reciprocity principle,

$$\frac{I_B}{V_A} = \frac{I_A}{V_B}$$

(8.43)

or, in other words, the ratio of the current induced in one antenna as a result of a voltage applied to the other is the same regardless of which antenna is used as the transmitter. Practically, this means that the design of the receiving antenna is just as important as the design of the transmitting antenna in optimizing the performance of an entire system.

References

Neiman, M.S. (1943). The principle of reciprocity in antenna theory. *Proceedings of the IRE* 31 (12): 666–671.

Stroud, K.A. (1996). *Further Engineering Mathematics: Programmes and Problems*. Basingstoke: Macmillan.

9

Electromagnetic Waves at Dielectric Interfaces

9.1 Introduction

In Chapter 7, we saw that it is possible to derive a wave equation for both electric and magnetic fields directly from the Maxwell equations as applied to dielectric media. This was a key result, as it provides a basis for understanding electromagnetic radiation in all its diversity, from very long-wavelength radio waves, though the visible spectrum to very short-wavelength X-rays. We have also seen how we can both transmit and receive electromagnetic waves that propagate through free space. We have not, however, considered what happens if an electromagnetic wave tries to pass from one dielectric medium to another.

Intuitively, we know that light may be reflected at the interface between two media – we see reflections on the surface of water, for example – and that this also applies across the electromagnetic spectrum. This is, after all, the basis of radar. We can emit a short radio-frequency pulse of electromagnetic radiation and measure the reflected signal, using the direction that the wave was propagating in and the time taken to receive the reflected signal to infer the position of an object. Understanding the behaviour of electromagnetic waves is also critical for the engineering of displays and photonic devices such as LEDs and lasers where light must be extracted from a dielectric medium to be of use.

In order to study the behaviour of electromagnetic waves at the interface between two dielectrics, we shall consider the simple generic situation shown in Figure 9.1 where we have two dielectric media which have a flat interface in the y–z plane at $x = 0$. The first medium extends to infinity in the negative x-direction and the second medium extends to infinity in the positive x-direction; the interface between the two is itself of infinite area. We will assume that the two dielectrics are non-magnetic and so both have a relative permeability of unity, but they have different relative permittivities of ε_{r1} and ε_{r2}, respectively.

A monochromatic plane electromagnetic wave with a Poynting vector $\mathbf{N}_i$ of the kind considered in Section 7.5 propagates through the first dielectric medium towards the interface where it is incident at an angle θ_i with respect to the line of the x-axis. Some of the electromagnetic wave will be reflected into the first dielectric medium with a Poynting vector $\mathbf{N}_r$ at an angle θ_r, while some will be transmitted into the second medium with a Poynting vector $\mathbf{N}_t$ at an angle θ_t.

Electromagnetism for Engineers, First Edition. Andrew J. Flewitt.
© 2023 John Wiley & Sons Ltd. Published 2023 by John Wiley & Sons Ltd.
Companion website: www.wiley.com/go/flewitt/electromagnetism

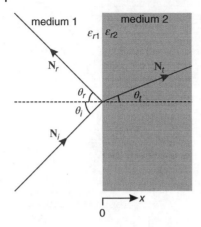

Figure 9.1 An incident monochromatic plane electromagnetic wave propagating through medium 1 is incident on a planar interface with medium 2. Some of the wave may be transmitted and some reflected.

Over the course of this chapter, we will consider what influences both the angle and intensity of the reflected and transmitted waves, but in order to be able to do this, we must first appreciate how electric and magnetic fields change across interfaces between dissimilar dielectrics.

9.2 Boundary Conditions

When we first considered dielectrics in Section 2.2, we saw that the application of an external electric field to a dielectric resulted in a surface polarization charge being formed, σ_P, per unit area. A consequence of this was that it was convenient to be able to define an electric flux density **D** in addition to the electric field **E**, where the two were related by

$$\mathbf{D} = \varepsilon_0 \varepsilon_r \mathbf{E} \tag{2.15}$$

In this way, the relative permittivity accounts for the effect of the surface polarization charge on the field inside the dielectric. However, we did not consider how the electric flux density and electric field change at the interface between two dielectric media. In order to understand this, let us consider the situation shown in Figure 9.2 where we have a block of some dielectric material suspended between the plates of an air-filled parallel plate capacitor. We can call the air 'dielectric 1' and the other material 'dielectric 2'. Dielectric 2 has been deliberately placed such that its surfaces are not parallel to the plates of the capacitor. This allows us to resolve both the flux density and field in each dielectric into components that are tangential to the dielectric interface, indicated with the subscript t, and components that are normal to the dielectric interface, indicated with the subscript n. For example, the electric flux density in dielectric 1 can be expressed as

$$\mathbf{D}_1 = \mathbf{D}_{n1} + \mathbf{D}_{t1} \tag{9.1a}$$

and in dielectric 2 as

$$\mathbf{D}_2 = \mathbf{D}_{n2} + \mathbf{D}_{t2} \tag{9.1b}$$

A layer of polarization charge will appear on the surface of the dielectric. To determine how this layer of charge affects the electric flux density across the interface, we will apply

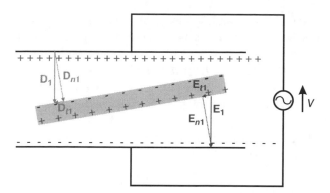

Figure 9.2 A two-dimensional representation of a block of dielectric material between the plates of a parallel plate capacitor. The electric flux density and electric field can each be resolved into components that are normal and tangential to the plane of the surface of the dielectric block.

Figure 9.3 A three-dimensional representation of the same situation as in Figure 9.2 where there is a block of dielectric material between the plates of a parallel plate capacitor. A Gaussian surface and closed loop are both indicated.

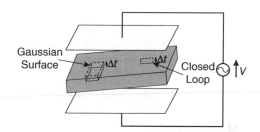

the Gauss law of electrostatics in integral form to a small cuboid Gaussian surface of infinitesimal thickness Δt and area A parallel to the interface, which straddles the interface as shown in Figure 9.3. We know that

$$\oint_S \mathbf{D.dA} = q_f \tag{2.14}$$

where q_f is the free charge enclosed by the surface. As the surface polarization charge is not a free charge, $q_f = 0$ for the surface in Figure 9.3. Therefore, as only the normal component of the electric flux density passes through this surface, Eq. (2.14) becomes

$$|\mathbf{D}_{n2}|A - |\mathbf{D}_{n1}|A = 0 \tag{9.2}$$

noting that the electric flux density is entering the Gaussian surface in dielectric 1 and exiting in dielectric 2, resulting in the different signs for the two components in Eq. (9.2). Therefore,

$$D_{n1} = D_{n2} \tag{9.3}$$

or, in other words, *the normal component of the electric flux density is conserved across a dielectric interface.*

Let us now consider the electric fields on either side of the interface using the Faraday law of magnetic fields in integral form, which states that

$$\oint_C \mathbf{E.dr} = -\int_S \frac{d\mathbf{B}}{dt}.\mathbf{dA} \tag{5.13}$$

where the left-hand side is the integral round a closed loop, such as that indicated in Figure 9.3, and the right-hand side is the integral over the surface bounded by the same loop. The specific loop shown in Figure 9.3 is aligned perpendicular to the plane of the dielectric surface with short sides of length Δt normal to the surface and long sides of length l tangential to the surface. If Δt is made to be very small, then the surface area bounded by the loop also becomes very small, and the right-hand side of Eq. (5.13) tends to zero. Meanwhile only the long sides are significant for the loop integral on the left-hand side of Eq. (5.13), for which the dot product reduces to a simple product of the component of the electric field tangential to the surface multiplied by the length of the side. Therefore, Eq. (5.13) reduces to

$$|\mathbf{E}_{t2}|l - |\mathbf{E}_{t1}|l = 0 \tag{9.4}$$

noting that the loop integration must be in the same direction all round the loop, and so the direction relative to the electric field will be opposite in the two materials resulting in the different signs again for the two components in Eq. (9.4). Therefore,

$$E_{t1} = E_{t2} \tag{9.5}$$

or in other words, *the tangential component of the electric field is conserved across a dielectric interface.*

We can use the same parallel plate capacitor system in Figures 9.2 and 9.3 to consider the boundary conditions for the magnetic flux density and magnetic field also. An a.c. voltage source is indicated in both figures, and therefore there will be a current flow around the simple capacitor circuit which will result in a magnetic field inside the current loop. We can again resolve the magnetic flux density and magnetic fields in each material into components that are normal and tangential to the surface using the same subscript notation as for the electric case. For example, the magnetic flux density in dielectric 1 can be expressed as

$$\mathbf{B}_1 = \mathbf{B}_{n1} + \mathbf{B}_{t1} \tag{9.6a}$$

and in dielectric 2 as

$$\mathbf{B}_2 = \mathbf{B}_{n2} + \mathbf{B}_{t2} \tag{9.6b}$$

Applying the Gauss law of magnetic fields (Eq. (3.22)) to the Gaussian surface shown in Figure 9.3 again results in only a normal component of any magnetic flux density passing through the surfaces of area A, resulting in a non-negligible flux such that

$$|\mathbf{B}_{n2}|A - |\mathbf{B}_{n1}|A = 0 \tag{9.7}$$

and hence,

$$B_{n1} = B_{n2} \tag{9.8}$$

In other words, *the normal component of the magnetic flux density is conserved across a dielectric interface.*

Finally, we can apply the Ampère–Maxwell law in integral form to the closed loop shown in Figure 9.3:

$$\oint_C \mathbf{H.dr} = I + \int_S \frac{\partial \mathbf{D}}{\partial t}.d\mathbf{A} \tag{5.34}$$

The key feature of this system is that there can be no surface currents at the dielectric interface as there are no free charges, and so $I = 0$ in Eq. (5.34). Furthermore, as the surface integral tends to zero if Δt is again made very small, the whole of the right-hand side of Eq. (5.34) tends to zero, and Eq. (5.34) reduces to

$$|\mathbf{H}_{t2}|l - |\mathbf{H}_{t1}|l = 0 \tag{9.9}$$

Therefore,

$$H_{t1} = H_{t2} \tag{9.10}$$

or in other words, *the tangential component of the magnetic field is conserved across a dielectric interface.*

We can therefore summarize these four conservation laws by saying that the normal component of the magnetic and electric flux densities and the tangential components of the electric and magnetic fields are conserved across dielectric boundaries, or in equations as

$$D_{n1} = D_{n2} \tag{9.11a}$$

$$B_{n1} = B_{n2} \tag{9.11b}$$

$$E_{t1} = E_{t2} \tag{9.11c}$$

$$H_{t1} = H_{t2} \tag{9.11d}$$

9.3 Angles of Reflection and Refraction

Figure 9.1 shows the basic situation of a monochromatic plane electromagnetic wave incident on a planar interface between two dielectric materials; a portion of this incident wave is transmitted into the medium and a portion is reflected. Therefore, there are three critical angles which must be related to each other: the angles of incidence θ_i, reflection θ_r and transmission θ_t (all of which are measured relative to a line perpendicular to the interface).

A simple way of understanding why these angles may not all be equal is to consider the wavefronts associated with each of the three waves. Wavefronts are the planes where the fields in the waves are a maximum, and they are therefore separated from each other by a wavelength λ. Figure 9.4 shows the wavefronts associated with the incident and transmitted plane electromagnetic waves. In this case, the incident wave covers a circular spot at the interface of diameter s. As the wave is monochromatic, there is a well-defined frequency f which is the same for all three waves. We know from Section 7.4 that the velocity of an electromagnetic wave is dependent on the dielectric medium through which it is passing. As the incident and reflected waves are in the same medium (1), they will both have the same wavelength λ_1 because the velocity of the waves is the same. However, if the permittivity of medium 2 is different than that of medium 1, then the velocity and hence the wavelength will also be different. We can express this quantitatively using the refractive index for the two media, n_1 and n_2 (see Section 7.4). From Eq. (7.16), we know that the velocity of the wave in medium 1 is

$$c_1 = \frac{c_{\text{vac}}}{n_1} \tag{9.12a}$$

where c_{vac} is the velocity of an electromagnetic wave in a vacuum. Similarly in medium 2,

$$c_2 = \frac{c_{vac}}{n_2} \tag{9.12b}$$

Therefore, using the standard relationship for waves, $c = f\lambda$, we can gain expressions for the wavelength in the two media from Eqs. (9.12) as

$$\lambda_1 = \frac{c_{vac}}{fn_1} \tag{9.13a}$$

$$\lambda_2 = \frac{c_{vac}}{fn_2} \tag{9.13b}$$

Figure 9.4 shows a moment in time when one incident wavefront has just arrived at one side of the circular spot at the interface and the diameter has been chosen such that at that moment a transmitted wavefront is just leaving the interface. In order to preserve the wavefronts as well-defined planes, it must take the same time for the wave to travel a distance λ_1 in medium 1 as it does to travel a distance λ_2 in medium 2. By looking at the geometry of the right-angle triangles formed by the wavefronts on either side of the interface, and remembering that the hypotenuse of both will be s, using the definition of sine we can see that

$$\frac{\lambda_1}{\sin \theta_i} = \frac{\lambda_2}{\sin \theta_t} \tag{9.14}$$

Substituting for the refractive index using Eq. (9.13) gives

$$n_1 \sin \theta_i = n_2 \sin \theta_t \tag{9.15}$$

The phenomenon of an electromagnetic wave changing direction when moving from one medium to another is known as *refraction*, and consequently Eq. (9.15) is commonly called the *Snell law of refraction*. It is for this reason that a straw standing in a glass of water appears to be distorted.

If a wave is passing from one dielectric material to another of higher refractive index ($n_2 > n_1$) then Eq. (9.15) means that the transmitted wave is refracted so that its path is closer to the normal ($\theta_t < \theta_i$); this is the situation shown in Figure 9.4.

However, if a wave is passing from one dielectric material to another of lower refractive index ($n_2 < n_1$) then the transmitted wave is refracted so that its path is further from the

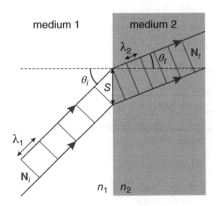

medium 1   medium 2

Figure 9.4 An incident monochromatic plane electromagnetic wave propagating through medium 1 is incident on a planar interface with medium 2; only the wave that is transmitted into the second medium is shown for clarity. The wavefronts associated with the incident and transmitted waves are shown as lines perpendicular to the direction of propagation.

normal ($\theta_t > \theta_i$). Clearly, θ_t cannot be greater than 90°. From Eq. (9.15), this situation will occur when

$$\theta_i > \sin^{-1}\left(\frac{n_2}{n_1}\right) \tag{9.16}$$

For these large angles of incidence there can be no transmitted wave, and instead the incident wave is all reflected back into medium 1. This *total internal reflection* is used to confine light to travel along optical fibres by fabricating the fibre from a central core dielectric with a higher refractive index than an outer dielectric cladding. We shall look at this in more detail in Section 11.7. It is also used in semiconductor lasers to confine photons in the lasing cavity.

In the case of the reflected wave, there is no change in medium compared to the incident wave and therefore no change in the velocity or wavelength either. As a result, in order to preserve the well-defined planes of the wavefronts in the reflected wave, it must be the case that the angle of reflection is the same as the angle of incidence,

$$\theta_r = \theta_i \tag{9.17}$$

which is what we intuitively know to be true.

9.4 The Fresnel Equations

Although basic geometry and consideration of the changing velocity of the electromagnetic wave can be used to determine the angle of the reflected and transmitted waves relative to a plane electromagnetic wave that is incident on an interface between two dielectric media, it does not give us quantitative information about the relative amplitude of the waves. This amplitude relationship is described by a set of equations that are collectively known as the *Fresnel equations*, which we shall derive in this section from the boundary conditions for the electric and magnetic fields and flux densities (see Section 9.2).

The boundary conditions in Eq. (9.11) set out which components of the electric and magnetic fields and flux densities must be conserved across a dielectric interface. In the general scenario described by Figure 9.1, we have two waves in medium 1: the incident and reflected waves. The principle of superposition means that these will sum at the interface, and the conserved components of these total fields and flux densities must equal those of the transmitted wave in medium 2. We can therefore generate equations which express this quantitatively. There is, however, one significant complexity: it is only the tangential component of the fields and the normal components of the flux densities that are conserved, but an arbitrary incident electromagnetic will have both normal and tangential components. Therefore, we have to consider separately two perpendicular polarizations of the incident electromagnetic wave: one where the electric field is perpendicular to the plane of incidence of the wave (Figure 9.5a) and one where it is parallel to the plane of incidence (Figure 9.5b), knowing that we can resolve any arbitrary incident wave into these two components (rather as we can resolve forces into two perpendicular components for the purposes of calculation). The plane of incidence is defined as the plane containing the Poynting vectors of the incident, reflected and transmitted waves, which is therefore the plane of the page of this book in Figure 9.5.

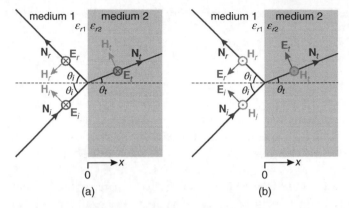

Figure 9.5 The two polarizations of the monochromatic plane electromagnetic wave shown in Figure 9.1: (a) with the electric field perpendicular to the plane of incidence (the plane of this page); (b) with the electric field parallel to the plane of incidence. In (a) the electric field is pointing away from you, while in (b) the magnetic field is pointing towards you.

We will first derive the Fresnel equations for the case where the electric field is perpendicular to the plane of incidence. The boundary condition of Eq. (9.11c) states that the component of the electric field tangential to a dielectric interface is conserved. It can be seen in Figure 9.5a that the electric fields in all of the incident, reflected and transmitted waves are tangential to the surface. The incident and reflected waves are pointing in the same direction, and so the simple sum of the magnitude of these electric fields must equal the electric field in the transmitted wave,

$$E_i + E_r = E_t \tag{9.18}$$

where the notation $E_i = |\mathbf{E}_i|$ is used. The tangential component of the magnetic field is also conserved across the dielectric interface (Eq. (9.11d)). However, it can be seen in Figure 9.5a that $\mathbf{H}_i$ has a component normal to the surface and a component tangential to the surface. Basic trigonometry allows the tangential component to be calculated from the angle of incidence as $H_i \cos \theta_i$. Similarly, the tangential component of the reflected wave is $H_r \cos \theta_i$ (remembering from Eq. (9.17) that the angles of incidence and reflection are equal), but this component is pointing in the opposite direction to that of the incident wave and so must be subtracted from it in determining the total tangential magnetic field. This will be equal to the tangential component of the transmitted magnetic field, $H_t \cos \theta_t$, giving

$$H_i \cos \theta_i - H_r \cos \theta_i = H_t \cos \theta_t \tag{9.19}$$

We will be able to combine Eqs. (9.18) and (9.19) if we re-express the latter in terms of electric fields using the intrinsic impedance (see Section 7.5) which is defined in Eq. (7.43) as being the ratio of the electric and magnetic fields and is a property of a given medium. Therefore, if the intrinsic impedances of medium 1 and medium 2 are η_1 and η_2 respectively, then for each of the incident, reflected and transmitted waves

$$H_i = E_i / \eta_1 \tag{9.20a}$$

$$H_r = E_r / \eta_1 \tag{9.20b}$$

$$H_t = E_t / \eta_2 \tag{9.20c}$$

noting that only the transmitted wave is in medium 2. Substituting Eq. (9.20) into Eq. (9.19) gives

$$\frac{E_i}{\eta_1} \cos \theta_i - \frac{E_r}{\eta_1} \cos \theta_i = \frac{E_t}{\eta_2} \cos \theta_t \qquad (9.21)$$

Equation (9.18) can now be used to substitute for E_r,

$$\frac{E_i}{\eta_1} \cos \theta_i - \frac{(E_t - E_i)}{\eta_1} \cos \theta_i = \frac{E_t}{\eta_2} \cos \theta_t \qquad (9.22)$$

which rearranges to

$$\left(\frac{E_t}{E_i} \right)_{E\perp} = \frac{2\eta_2 \cos \theta_i}{\eta_2 \cos \theta_i + \eta_1 \cos \theta_t} \qquad (9.23a)$$

Alternatively, Eq. (9.18) can be used to substitute for E_t in Eq. (9.21) to give

$$\left(\frac{E_r}{E_i} \right)_{E\perp} = \frac{\eta_2 \cos \theta_i - \eta_1 \cos \theta_t}{\eta_2 \cos \theta_i + \eta_1 \cos \theta_t} \qquad (9.23b)$$

Eqs. (9.23a) and (9.23b) are the first pair of Fresnel equations; the subscript $E\perp$ reminds us that they only apply when the electric field is perpendicular to the plane of incidence. Specifically, they quantify the ratios of the electric field amplitude of the transmitted wave to the incident wave and the reflected wave to the incident wave, respectively.

We can derive a similar pair of Fresnel equations for the case where the electric field is perpendicular to the plane of incidence, as shown in Figure 9.5b. In this case, only a component of the electric fields in the three waves is tangential to the surface, and indeed the tangential component of the reflected wave is pointing in the opposite direction to that of the incident wave so that the former must be subtracted from the latter to give the total tangential electric field in medium 1 at the interface. This is equal to the tangential component of the transmitted wave,

$$E_i \cos \theta_i - E_r \cos \theta_i = E_t \cos \theta_t \qquad (9.24)$$

Meanwhile, this time the magnetic fields in all of the incident, reflected and transmitted waves are tangential to the surface with the incident and reflected waves pointing in the same direction. Therefore, applying the conservation of the tangential magnetic field across the interface gives

$$H_i + H_r = H_t \qquad (9.25)$$

Once again, we can use the intrinsic impedance relations in Eq. (9.20) to re-express Eq. (9.25) in terms of electric fields,

$$\frac{E_i}{\eta_1} + \frac{E_r}{\eta_1} = \frac{E_t}{\eta_2} \qquad (9.26)$$

Finally, substitution for E_r from Eq. (9.26) into Eq. (9.24) gives

$$\left(\frac{E_t}{E_i} \right)_{E\parallel} = \frac{2\eta_2 \cos \theta_i}{\eta_1 \cos \theta_i + \eta_2 \cos \theta_t} \qquad (9.27a)$$

while substituting for E_t from Eq. (9.26) into Eq. (9.24) gives

$$\left(\frac{E_r}{E_i} \right)_{E\parallel} = \frac{\eta_1 \cos \theta_i - \eta_2 \cos \theta_t}{\eta_1 \cos \theta_i + \eta_2 \cos \theta_t} \qquad (9.27b)$$

The subscript $E\|$ on this pair of Fresnel equations reminds us that they only apply when the electric field is perpendicular to the plane of incidence.

There are a few points of note regarding the two pairs of Fresnel equations (Eqs. (9.24) and (9.27)). The first is their similarity to the expressions for the voltage reflection and transmission coefficients for guided electromagnetic waves on transmission lines (Eqs. (6.68) and (6.81), respectively). In all cases, the ratio of the transmitted wave to the incident wave is some function of twice the impedance of the 'load' divided by a quantity related to summing components of the 'input' and 'load' (where in this case medium 1 is the 'input' and medium 2 is the 'load'). Meanwhile, the ratio of the reflected wave to the incident wave is a quantity related to the difference between the 'input' and 'load' divided by a quantity related to summing components of the 'input' and 'load'. This similarity becomes even more apparent when the angle of incidence $\theta_i = 0$, as we shall see when we look at anti-reflection coatings in Section 9.6. This similarity is because the transmission line is really just a special case of guiding an electromagnetic wave through a dielectric medium. Therefore, the same underlying physics applies when we move from one medium to another in both transmission lines and free-space electromagnetic waves. Secondly, we saw in transmission lines that the voltage of an incident wave at a connection between transmission lines is not equal to the simple sum of the voltages of the reflected and transmitted waves as current has to be conserved and the characteristic impedance of the two transmission lines is different. In the case of the free-space wave, power has to be conserved and the intrinsic impedance of the two media is not the same, and therefore the electric field of the incident wave is not the simple sum of the electric fields of the reflected and transmitted waves.

Finally, we should remember that the electric and magnetic fields are related to each other by the intrinsic impedance (Eq. (9.20)). As the reflected and incident waves are in the same medium,

$$\frac{H_r}{H_i} = \frac{E_r}{E_i} \tag{9.28}$$

and therefore Eqs. (9.23b) and (9.27b) can both be immediately rewritten in terms of magnetic fields as

$$\left(\frac{H_r}{H_i}\right)_{E\perp} = \frac{\eta_2 \cos\theta_i - \eta_1 \cos\theta_t}{\eta_2 \cos\theta_i + \eta_1 \cos\theta_t} \tag{9.29a}$$

$$\left(\frac{H_r}{H_i}\right)_{E\|} = \frac{\eta_1 \cos\theta_i - \eta_2 \cos\theta_t}{\eta_1 \cos\theta_i + \eta_2 \cos\theta_t} \tag{9.29b}$$

However, the transmitted and incident waves are in different media, so from Eq. (9.20)

$$\frac{H_t}{H_i} = \frac{\eta_1 E_t}{\eta_2 E_i} \tag{9.30}$$

Therefore, Eqs. (9.23a) and (9.27a) are not the same when rewritten in terms of magnetic fields, but become

$$\left(\frac{H_t}{H_i}\right)_{E\perp} = \frac{2\eta_1 \cos\theta_i}{\eta_2 \cos\theta_i + \eta_1 \cos\theta_t} \tag{9.31a}$$

$$\left(\frac{H_t}{H_i}\right)_{E\|} = \frac{2\eta_1 \cos\theta_i}{\eta_1 \cos\theta_i + \eta_2 \cos\theta_t} \tag{9.31b}$$

9.5 Polarization by Reflection

A key result from the Fresnel equations is that the component of an arbitrary electromagnetic wave with the electric field parallel to the plane of incidence may be reflected differently than the component with the electric field perpendicular to the plane of incidence. Indeed, there is a special case when $E_r/E_i = 0$ for the parallel polarization only. We can see this by plotting the ratio of the reflected electric field to the incident electric field as a function of angle of incidence for the situation of light travelling through air ($\eta = 377\,\Omega$) incident on water ($\eta = 42\,\Omega$) as shown in Figure 9.6. At an angle ~84°, only the wave with the electric field component perpendicular to the plane of incidence is reflected. The angle at which this occurs is known as the *Brewster angle*. Although the incident wave is not polarized, the reflected wave is plane polarized, and therefore this phenomenon is known as *polarization by reflection*.

Polarization by reflection only occurs at a specific Brewster angle for the situation where the intrinsic impedance of the medium that is being entered is less than that of the medium that is being left ($\eta_2 < \eta_1$) which means that the angle of transmission must be less than the angle of incidence ($\theta_t < \theta_i$) according to the Snell law (Eq. (9.15)). We should also remember that a higher intrinsic impedance means a lower refractive index. If we consider the equation for reflection of the perpendicular component (Eq. (9.23b)), we can see that the numerator is $\eta_2 \cos\theta_i - \eta_1 \cos\theta_t$. Therefore, if $\theta_t < \theta_i$ then $\cos\theta_t > \cos\theta_i$ as well as $\eta_1 > \eta_2$ and so it must always be true that $\eta_1 \cos\theta_t > \eta_2 \cos\theta_i$, and E_r/E_i cannot be zero. In other words, there will always be some reflection of the perpendicular component of the electromagnetic wave. However, the equation for the reflection of the parallel component (Eq. (9.27b)) has $\eta_1 \cos\theta_i - \eta_2 \cos\theta_t$ for the numerator. With the same conditions for the

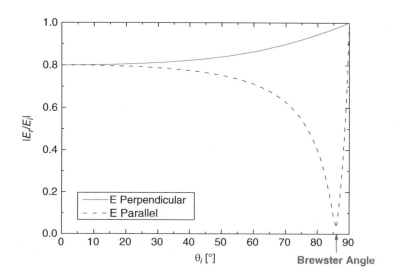

Figure 9.6 The ratio of the incident to reflected electric field for the parallel and perpendicular electric field polarization for a wave travelling from air into water as a function of angle of incidence, showing clearly the Brewster angle where only the perpendicular polarization is reflected.

relative values of the terms as already described, it is entirely possible for the situation where $\eta_1 \cos\theta_i = \eta_2 \cos\theta_t$ to occur, so that $E_r/E_i = 0$. This is what is observed in Figure 9.6.

In order to calculate the Brewster angle θ_B (the angle of incidence when the reflected wave is polarized) we need to combine the condition for no reflection of the parallel component when the numerator of Eq. (9.27b) is zero,

$$\eta_1 \cos\theta_i = \eta_2 \cos\theta_t \tag{9.32}$$

with the Snell law (Eq. (9.15)). However, the latter is written in terms of refractive index rather than intrinsic impedance. We know that the characteristic impedance of a medium is given by

$$\eta = \sqrt{\frac{\mu_0 \mu_r}{\varepsilon_0 \varepsilon_r}} \tag{7.43}$$

but also that, for non-magnetic dielectrics

$$n = \sqrt{\varepsilon_r} \tag{7.18}$$

and therefore $n \propto 1/\eta$. The Snell law (Eq. (9.15)) can therefore be rewritten in terms of intrinsic impedances as

$$\eta_2 \sin\theta_i = \eta_1 \sin\theta_t \tag{9.33}$$

This can be rearranged to

$$\sin^2\theta_t = \left(\frac{\eta_2}{\eta_1}\right)^2 \sin^2\theta_i \tag{9.34}$$

Similarly, we can arrange the condition for no parallel reflection (Eq. (9.32)) to

$$\cos^2\theta_t = \left(\frac{\eta_1}{\eta_2}\right)^2 \cos^2\theta_i \tag{9.35}$$

However, from basic trigonometry

$$\sin^2\theta_t + \cos^2\theta_t = 1 = \sin^2\theta_i + \cos^2\theta_i \tag{9.36}$$

and so we can combine Eqs. (9.34) and (9.35) by substitution into Eq. (9.36) to eliminate the transmission angle

$$\left(\frac{\eta_2}{\eta_1}\right)^2 \sin^2\theta_i + \left(\frac{\eta_1}{\eta_2}\right)^2 \cos^2\theta_i = \sin^2\theta_i + \cos^2\theta_i \tag{9.37}$$

This can now be rearranged to

$$\left(\frac{\eta_2^2 - \eta_1^2}{\eta_1^2}\right) \sin^2\theta_i = \left(\frac{\eta_2^2 - \eta_1^2}{\eta_2^2}\right) \cos^2\theta_i \tag{9.38}$$

$$\tan^2\theta_i = \left(\frac{\eta_1^2}{\eta_2^2}\right) \tag{9.39}$$

Therefore, knowing that this angle of incidence is the Brewster angle (i.e. $\theta_B = \theta_i$) gives the final result:

$$\theta_B = \tan^{-1}\left(\frac{\eta_1}{\eta_2}\right) \tag{9.40}$$

Sunglasses are perhaps the more common application of polarization by reflection in everyday life. The purpose of sunglasses is primarily to reduce the amount of sunlight reflected from a surface (particularly highly reflective surfaces such as water, snow, sand or roads) that reaches the eyes. In all these cases, the material which the sunlight is being reflected off has a lower intrinsic impedance than air, and therefore the component of light with the electric field perpendicular to the plane of incidence is reflected with a much higher amplitude than the parallel component. Therefore, sunglasses include a polaroid which is oriented to absorb the perpendicular component of the light, significantly reducing the glare from the surface. The glasses are manufactured assuming that the wearer is upright and the head is level (which is usually the case). However, if you stand facing the sun wearing sunglasses with a highly reflective surface in front of you (such as a road) and you tilt your head to one side, you will see a significant increase in the reflected light intensity as you start to allow the perpendicular component to pass through the polaroid.

9.6 Anti-reflection Coatings

We have seen that an electromagnetic wave undergoes some degree of reflection when it is incident upon an interface between two materials of different characteristic impedance. This presents a problem as there are many situations where we need an electromagnetic wave to travel between different media with minimum loss of power, such as the coupling of sunlight into a solar cell, the transfer of light into and out of an optical fibre, or even the passage of light through a pair of glasses. We faced a similar issue with transmission lines where we needed to couple a line with one characteristic impedance to a load of a different impedance (see Section 6.8). In that case, we found that we could perform quarter-wave matching to eliminate this problem: the addition of a quarter wavelength of transmission line with a characteristic impedance equal to the geometric mean of the incoming line and the load. We shall see that we can use a similar approach in this new case by adding a thin film of material at the interface between the two materials which we wish our electromagnetic wave to pass between, called an *anti-reflection coating*.

In order to understand what the properties of this anti-reflection coating should be, we need to simplify the problem by considering the case where the incident wave is perpendicular to the plane of incidence, so that $\theta_i = 0°$ as shown in Figure 9.7. The wave is therefore travelling in the x-direction, and we continue to define $x = 0$ as being at the interface. At this point, we face a mathematical oddity in that the incident, reflected and transmitted waves are all travelling along the same line, and so there is no single plane of incidence as an infinite number of planes can pass through a line. However, for the purposes of being able to perform a calculation in Cartesian coordinates, we still need to define a plane of incidence that we can use to resolve the wave into components where the electric field is parallel and perpendicular to this plane. In the end, we shall see that this has no effect on the final result, as we should expect.

The first step in understanding how to engineer an anti-reflection coating that will minimize reflections at an interface is to calculate the effective impedance in the medium of the incident wave at any distance x away from the interface. The intrinsic impedance of a medium is the ratio of the electric and magnetic fields for a unidirectional wave in the

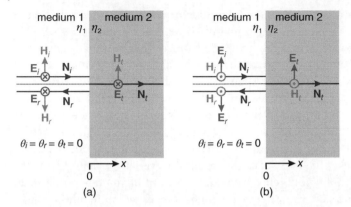

Figure 9.7 Schematic representation of the two possible polarizations of the incident electromagnetic wave relative to an (arbitrary) plane of incidence which is the plane of the page. (a) The electric field is perpendicular to the plane of incidence. (b) The electric field is parallel to the plane of incidence.

absence of reflections. However, in this case we do have reflections, and so the apparent impedance is the ratio of the total electric and magnetic fields. For the situation where the electric field is perpendicular to the plane of incidence, as shown in Figure 9.7a, the electric fields of the incident and reflected waves point in the same direction and therefore sum, but the magnetic field points in opposite directions and so subtract such that

$$\eta(x) = \frac{\overline{E_t}e^{j(\omega t - \beta x)} + \overline{E_r}e^{j(\omega t + \beta x)}}{\overline{H_t}e^{j(\omega t - \beta x)} - \overline{H_r}e^{j(\omega t + \beta x)}} \tag{9.41}$$

where $\overline{E_t}$ and $\overline{E_r}$ are the electric field's amplitude and phase of the incident and reflected waves respectively at $x = 0$ and $\overline{H_t}$ and $\overline{H_r}$ are the similar expressions for the magnetic fields (note the similarity with Eqs. (6.28) and (6.29) for transmission lines). The common factor of $e^{j\omega t}$ can be cancelled out from all terms in Eq. (9.41), and we can use the intrinsic impedance of the medium

$$\eta_1 = \frac{E_i}{H_i} = \frac{E_r}{H_r} \tag{9.42}$$

to substitute for the magnetic field terms to give

$$\eta(x) = \frac{\overline{E_t}e^{-j\beta x} + \overline{E_r}e^{j\beta x}}{\frac{\overline{E_t}}{\eta_1}e^{-j\beta x} - \frac{\overline{E_r}}{\eta_1}e^{j\beta x}} = \eta_1 \left[\frac{e^{-j\beta x} + \left(\frac{E_r}{E_i}\right)_{E\perp} e^{j\beta x}}{e^{-j\beta x} - \left(\frac{E_r}{E_i}\right)_{E\perp} e^{j\beta x}} \right] \tag{9.43}$$

In this way, we have been able to express the apparent impedance in terms of the Fresnel equations that we derived in Section 9.4. Specifically, we can substitute Eq. (9.23b) into Eq. (9.43), but with the simplification that $\theta_i = \theta_t = 0$, so that all the cosine terms become unity such that

$$\eta(x) = \eta_1 \left[\frac{e^{-j\beta x} + \frac{\eta_2 - \eta_1}{\eta_2 + \eta_1}e^{j\beta x}}{e^{-j\beta x} - \frac{\eta_2 - \eta_1}{\eta_2 + \eta_1}e^{j\beta x}} \right] = \eta_1 \left[\frac{(\eta_2 + \eta_1)e^{-j\beta x} + (\eta_2 - \eta_1)e^{j\beta x}}{(\eta_2 + \eta_1)e^{-j\beta x} - (\eta_2 - \eta_1)e^{j\beta x}} \right] \tag{9.44}$$

This equation is completely analogous to that for the apparent impedance a distance x away from a load on a transmission line (Eq. (6.86)). Once again, it is easier to see what is happening by using the de Moivre theorem to express Eq. (9.44) in terms of trigonometric functions, giving

$$\eta(x) = \eta_1 \left[\frac{\eta_2 - j\eta_1 \tan(\beta x)}{\eta_1 - j\eta_2 \tan(\beta x)} \right] \tag{9.45}$$

The apparent impedance is therefore a smoothly varying function of x. At the interface between the two dielectrics, where $x = 0$, $\tan(\beta x) = 0$ and so the apparent impedance becomes that of the dielectric that the electromagnetic wave is passing into (η_2). However, there is a further special case when $x = -\lambda/4$, as at this distance from the interface $\beta x = -\pi/2$ and so $\tan(\beta x) = -\infty$. Equation (9.45) then becomes

$$\eta\left(x = -\frac{\lambda}{4}\right) = \frac{\eta_1^2}{\eta_2} \tag{9.46}$$

To understand how we can use this result in practice, let us imagine that we want to allow an electromagnetic wave, for example monochromatic laser light, to pass from one dielectric medium with an intrinsic impedance of η_0, such as air, into a second dielectric medium of intrinsic impedance η_2 which might be an optical fibre made from glass. If we coat the optical fibre with a layer of another dielectric medium with an intrinsic impedance η_1 equal to the geometric mean of the other two media,

$$\eta_1 = \sqrt{\eta_0 \eta_2} \tag{9.47}$$

and a thickness equal to a quarter of the wavelength of the electromagnetic wave, as shown in Figure 9.8, then the apparent impedance that the incoming wave will 'see' when it enters the additional layer of material will be η_0. As there is no change in apparent impedance, there will be no reflection of the wave. Inside the additional layer, the impedance will slowly change until it is η_2 when it reaches the other dielectric. Again, there is no change in apparent impedance, and no wave will be reflected. The result of adding this layer of material, an *anti-reflection coating*, is that all the incoming wave (from the air in the specific example) is transmitted into the second dielectric (the optical fibre).

There are two ways of thinking about how the quarter-wavelength anti-reflection coating works. One is the purely mathematical approach that we have used so far, but the

Figure 9.8 The application of an anti-reflection coating of a quarter-wavelength thickness results in the primary reflection being exactly cancelled by the secondary reflections due to destructive interference. The situation where $\eta_0 > \eta_1 > \eta_2$ is shown.

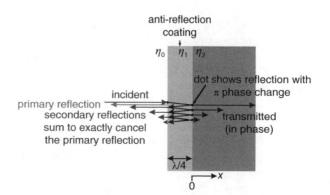

other is to consider how an incident wave will be reflected by the multiple layers of materials. Let us consider the case shown in Figure 9.8 where the three materials have intrinsic impedances such that $\eta_0 > \eta_1 > \eta_2$ (remembering that as η_1 is the geometric mean of the other two materials, the only other possible scenario is that $\eta_0 < \eta_1 < \eta_2$). This would be true for the scenario of the laser light passing from air into the optical fibre. When the light reaches the interface between the air and the anti-reflection coating, there is a difference in intrinsic impedance, and so a proportion of the wave will be reflected, given by Eq. (9.23b). We will call this the primary reflection, and as $\eta_0 > \eta_1$, the reflection coefficient is negative; it undergoes a π phase shift with respect to the wave which enters the anti-reflection coating. This transmitted wave will then reach the interface between the anti-reflection coating and the optical fibre. Some of the wave will be transmitted, but some will be reflected back into the anti-reflection coating and as $\eta_1 > \eta_2$ there is another π phase shift, and this reflected wave will get back to the interface between the anti-reflection coating and the air where some of the wave will be transmitted back out into the air. We will call this wave, which has travelled through the anti-reflection coating, a secondary reflection. Both the primary and secondary reflections have had π phase shifts, but the secondary reflection has travelled an extra $\lambda/2$ distance, and so the two are in antiphase and will destructively interfere with each other. Part of the wave will be reflected into the anti-reflection coating, but this time there is no phase change. It will continue to 'bounce' backwards and forwards inside the anti-reflection coating many times, transmitting some wave into the optical fibre and some back into the air as a secondary reflection each time. However, for every journey across the anti-reflection coating, the wave travels an extra $\lambda/2$ distance and undergoes a π phase change, and so all of the secondary reflections are in phase with each other; it is the requirement that the anti-reflection coating is $\lambda/4$ thick that ensures this is the case. However, the other condition that the intrinsic impedance of the anti-reflection coating is $\sqrt{\eta_0\eta_2}$ ensures that the sum of all of the secondary reflections exactly cancels with the primary reflection, with the result that there is no reflected wave. All of the power in the wave must therefore be transmitted into the glass.

Quarter-wavelength anti-reflection coatings are widely used in optical systems, as typical thicknesses are in the range of around 50–1000 nm, and it is relatively easy to coat surfaces in dielectric layers of these thicknesses. Furthermore, in some systems we are using monochromatic light of a well-defined wavelength and we can ensure that the light is incident at an angle of zero, making design of a highly efficient coating relatively easy. An example of this is the laser beam coupling into the optical fibre. It is more challenging if we are trying to couple sunlight into a solar cell. Not only do we have to cope with the sunlight having a wide range of wavelengths, but unless we can continuously move the orientation of the cell during the day, it is unlikely that the light will be at normal incidence. We will never be able to engineer a perfect anti-reflection coating under these circumstances – we just have to try and minimize reflections. Nature therefore sometimes takes a different approach. The eye of a moth is coated in cone-like nanostructures which are much smaller than the wavelength of light. The result is that incoming light of all wavelengths and angles of incidence sees a relatively smooth variation in intrinsic impedance as it passes through this small forest of nanostructures. In this way the moth, which has to live in low-light conditions, maximizes the light entering its eye. Such approaches have been suggested to improve the efficiency of solar cells (Diedenhofen et al. 2009).

Finally, in deriving the properties of the anti-reflection coating, we assumed the situation where the electric field is perpendicular to the plane of incidence, which gave Eq. (9.41). If the electric field is parallel to our (very arbitrary) plane of incidence, as shown in Figure 9.7b, then the equivalent equation for the effective impedance is

$$\eta(x) = \frac{\overline{E_i}e^{j(\omega t - \beta x)} - \overline{E_r}e^{j(\omega t + \beta x)}}{\overline{H_i}e^{j(\omega t - \beta x)} + \overline{H_r}e^{j(\omega t + \beta x)}} \tag{9.48}$$

as this time it is the electric fields that subtract and the magnetic fields that sum. This simplifies to

$$\eta(x) = \eta_1 \left[\frac{e^{-j\beta x} - \left(\frac{E_r}{E_i}\right)_{E\parallel} e^{j\beta x}}{e^{-j\beta x} + \left(\frac{E_r}{E_i}\right)_{E\parallel} e^{j\beta x}} \right] \tag{9.49}$$

We now substitute in Eq. (9.27b) for the proportion of the electric field that is reflected,

$$\eta(x) = \eta_1 \left[\frac{e^{-j\beta x} - \frac{\eta_1 - \eta_2}{\eta_1 + \eta_2} e^{j\beta x}}{e^{-j\beta x} + \frac{\eta_1 - \eta_2}{\eta_1 + \eta_2} e^{j\beta x}} \right] = \eta_1 \left[\frac{(\eta_1 + \eta_2)e^{-j\beta x} + (\eta_2 - \eta_1)e^{j\beta x}}{(\eta_1 + \eta_2)e^{-j\beta x} - (\eta_2 - \eta_1)e^{j\beta x}} \right] \tag{9.50}$$

which is identical to Eq. (9.44). Therefore it really does not matter what the polarization of the wave is; the behaviour of the anti-reflection coating is always the same.

Reference

Diedenhofen, S.L., Vecchi, G. et al. (2009). Broad-band and omnidirectional antireflection coatings based on semiconductor nanorods. *Advanced Materials* 21 (9): 973–978.

10

Electromagnetic Waves in Conducting Media

10.1 The Maxwell Equations in Conducting Media

In Chapter 7, we looked at how electric and magnetic fields behave in dielectric media and, using the Maxwell equations, we were able to prove that electromagnetic waves can propagate through these materials. However, we intuitively know that the situation must be very different in conductors. For a start, metals act as very good reflectors of electromagnetic waves; anyone who has tried to get a mobile phone signal inside a metal-framed building will be familiar with this effect, and we make mirrors out of thin metallic films.

The origin of this profound difference between how electromagnetic waves interact with dielectrics and with conductors lies in the presence of mobile electrons which can easily move under the influence of an applied electric field. We discussed this previously in Section 2.6 for metals, concluding that the consequence was that static electric fields cannot penetrate into these materials, as the free electrons can always cause a surface charge to form on a good conductor to counteract the effect of the applied field. However, it must take some finite time for the mobile electrons to form such a surface charge, and therefore it is less clear whether a high-frequency oscillating electric field, as found in an electromagnetic wave, would be able to penetrate into a conductor. It turns out that such an intuitive view is indeed correct and, as usual, we can prove this using the Maxwell equations.

The Maxwell equations, which are summarized in Table 5.1, apply to all media, including conductors, and just as we were able to simplify them in dielectrics, so we can also simplify them in conductors. In dielectrics, the lack of any mobile, free charges means that there clearly cannot be a current density ($\mathbf{J} = 0$) and also there is no net charge density in any volume of space as charge cannot move to cause an imbalance between positive and negative charge ($\rho = 0$). In conductors, we do have free, mobile charge and therefore there may be a current density $\mathbf{J}$. However, it remains true that the charge density in any volume of space is zero, as any non-zero charge density would cause a local electric field according to the Gauss law of electric fields (Eq. (5.7)) which would then cause the free charge to move until there is zero charge density again so that there is no electric field. Setting $\rho = 0$ means that the Maxwell equations in conductors may be expressed as

$$\nabla . \mathbf{D} = 0 \tag{10.1a}$$

$$\nabla . \mathbf{B} = 0 \tag{10.1b}$$

Electromagnetism for Engineers, First Edition. Andrew J. Flewitt.
© 2023 John Wiley & Sons Ltd. Published 2023 by John Wiley & Sons Ltd.
Companion website: www.wiley.com/go/flewitt/electromagnetism

$$\nabla \times \mathbf{E} = -\frac{\partial \mathbf{B}}{\partial t} \tag{10.1c}$$

$$\nabla \times \mathbf{H} = \mathbf{J} + \frac{\partial \mathbf{D}}{\partial t} \tag{10.1d}$$

In practice, it is only the Gauss law of electric fields (Eq. (10.1a)) that is changed by this simplification for conductors from the completely generalized form of the Maxwell equations. This means that there are no electric monopoles in the bulk of a conductor, and therefore lines of electric flux must form closed loops. Comparison with the Maxwell equations in dielectrics (Eq. (7.1)) reveals that the only difference is the inclusion of the current density term in Eq. (10.1d) for the Ampère–Maxwell law in conductors. Although this might seem like a minor difference, which is saying that a moving charge in addition to a changing electric flux density could produce a circulating magnetic field, we shall see that it has a profound effect upon the behaviour of electromagnetic waves.

10.2 The Electromagnetic Wave Equation in Conducting Media

In Section 7.3, we saw that it is possible to combine the Maxwell equations, each of which is only expressing mathematically features of the origin or nature of electric and magnetic fields, to create a wave equation. This not only proved the existence of electromagnetic waves, but also predicted specific details about their behaviour in dielectrics, such as their speed and the relative magnitude and orientation of the electric and magnetic fields. We can follow exactly the same process to create a wave equation for electromagnetic waves in conductors.

Our starting point again is to take the curl of the Faraday law of magnetic fields (Eq. (10.1c)) to give

$$\nabla \times \nabla \times \mathbf{E} = -\frac{\partial}{\partial t}(\nabla \times \mathbf{B}) \tag{10.2}$$

and to use the vector calculus identity that for an arbitrary vector quantity $\mathbf{Y}$,

$$\nabla \times \nabla \times \mathbf{Y} = \nabla(\nabla.\mathbf{Y}) - \nabla^2 \mathbf{Y} \tag{10.3}$$

with the result that Eq. (10.2) can be rewritten as

$$\nabla(\nabla.\mathbf{E}) - \nabla^2 \mathbf{E} = -\frac{\partial}{\partial t}(\nabla \times \mathbf{B}) \tag{10.4}$$

Also, as $\mathbf{D} = \varepsilon_0 \varepsilon_r \mathbf{E}$ (Eq. (2.15)), it follows from the Gauss law of electric fields (Eq. (10.1a)) that $\nabla.\mathbf{E} = 0$ and Eq. (10.4) reduces to

$$\nabla^2 \mathbf{E} = \frac{\partial}{\partial t}(\nabla \times \mathbf{B}) \tag{10.5}$$

Assuming that the conductor is behaving as a simple ideal linear magnetic material (see Section 4.2), then $\mathbf{B} = \mu_0 \mu_r \mathbf{H}$ and Eq. (10.5) becomes

$$\nabla^2 \mathbf{E} = \mu_0 \mu_r \frac{\partial}{\partial t}(\nabla \times \mathbf{H}) \tag{10.6}$$

Up to this point, the derivation of the wave equation in a conductor is identical to the dielectric case (see Section 7.3), but now we need to substitute for the Ampère–Maxwell law (Eq. (10.1d)) which has the additional current density term,

$$\nabla^2 \mathbf{E} = \mu_0 \mu_r \frac{\partial}{\partial t} \left(\mathbf{J} + \frac{\partial \mathbf{D}}{\partial t} \right) \tag{10.7}$$

Our aim remains the same as in the previous derivation: to create an equation entirely in terms of the electric field. The electric flux density term in Eq. (10.7) can easily be converted into the electric field as the two are related by the permittivity of the medium $\mathbf{D} = \varepsilon_0 \varepsilon_r \mathbf{E}$. Meanwhile, the Ohm law, which we are probably most familiar with in the form $V = IR$, where V is the voltage across an object, I is the current flowing through the object and R is the object's resistance, can be rewritten in terms of the basic conductivity σ of the material from which the object is made as

$$\mathbf{J} = \sigma \mathbf{E} \tag{10.8}$$

This form of the Ohm law has the advantage that it is independent of the specific geometry of the object through which charge is flowing. These two substitutions allow us to express Eq. (10.7) as

$$\nabla^2 \mathbf{E} = \mu_0 \mu_r \frac{\partial}{\partial t} \left(\sigma \mathbf{E} + \varepsilon_0 \varepsilon_r \frac{\partial \mathbf{E}}{\partial t} \right)$$
$$= \mu_0 \mu_r \sigma \frac{\partial \mathbf{E}}{\partial t} + \mu_0 \mu_r \varepsilon_0 \varepsilon_r \frac{\partial^2 \mathbf{E}}{\partial t^2} \tag{10.9}$$

This is another three-dimensional wave equation which is identical to that derived for dielectrics (Eq. (7.7)) except that the additional current density term in the Ampère–Maxwell law has led to an additional term in Eq. (10.9) which is dependent on the first differential of the electric field with respect to time.

We now need to find a possible solution to our new wave equation. Just as for the simple dielectric case, we will consider a monochromatic plane wave. The significance and properties of such waves were explained previously in Section 7.5, but the key fact is that any complex wave in three dimensions can be expressed as the sum of an infinite series of plane waves using Fourier analysis, and so the insights that we gain using this simple wave allow us to understand more complex scenarios as well.

The general solution to Eq. (10.9) for a monochromatic plane wave propagating in the positive z-direction is

$$\mathbf{E} = (E_{0x}\mathbf{i} + E_{0y}\mathbf{j} + E_{0z}\mathbf{k}) \exp(j\omega t) \exp(-\gamma z) \tag{10.10}$$

where $\mathbf{i}, \mathbf{j}$ and $\mathbf{k}$ are the unit vectors, E_{0x}, E_{0y}, and E_{0z} are the components of the amplitude of the electric field in each of the x-, y- and z-directions respectively, and γ is a new term called the *propagation constant*. It is a complex number of the form

$$\gamma = \alpha + j\beta \tag{10.11}$$

where α and β are both real numbers whose physical significance we shall shortly understand in Section 10.3.

When we studied the properties of electromagnetic waves in dielectrics, we were able to consider the simplified case of linearly polarized monochromatic plane waves (see Section 7.5). This was because the absence of net charge in the bulk of a dielectric meant that the

Gauss law of electric fields is $\nabla \cdot \mathbf{D} = 0$, and therefore there could be no component of the electric field in the direction of propagation of the wave. As it is also true that there is no net charge in the bulk of a conductor, the Gauss law of electric fields is unchanged (Eq. (10.1a)) and so it remains the case that there is no component of electric field in the direction of propagation of the electromagnetic wave: $E_{0z} = 0$ in Eq. (10.10). Therefore, we shall consider the special case where the electric field is pointing in the x-direction (i.e. $E_{0y} = 0$), knowing that we could superpose a second linearly polarized wave where the electric field is pointing in the y-direction to create any arbitrary wave. Such a linearly polarized monochromatic plane electromagnetic wave based on Eq. (10.10) will have the form

$$\mathbf{E} = E_{0x}\mathbf{i} \exp(j\omega t) \exp(-\gamma z) \tag{10.12}$$

We can now substitute the complex form of the propagation constant (Eq. (10.11)) into the expression for the linearly polarized monochromatic plane electromagnetic wave and group the spatial and temporal terms to give

$$\mathbf{E} = E_{0x}\mathbf{i} \exp(j[\omega t - \beta z]) \exp(-\alpha z) \tag{10.13}$$

In many ways, we have arrived at a solution which is almost identical to the linearly polarized electromagnetic wave in dielectrics (Eq. (7.34)). The amplitude of the electric field is E_{0x} at $z = 0$, the unit vector $\mathbf{i}$ means that this field points in the x-direction, and the $\exp(j[\omega t - \beta z])$ term means that it is propagating in the positive z-direction with an angular frequency of ω and a wavelength $\lambda = 2\pi/\beta$. However, the new $\exp(-\alpha z)$ term means that the amplitude is decreasing exponentially with distance, as shown in Figure 10.1.

If the wave amplitude is decreasing as the wave propagates through the conductor then the energy in the wave must be being dissipated in the conductor. This is because the conductor must contain mobile charge and so the oscillating electric field will cause a localized oscillating current to flow. This will result in energy dissipation in the form of heat as the conductor must have a finite resistivity, and this energy must be coming from the electromagnetic wave. We shall consider this in more quantitative detail in Section 10.3.

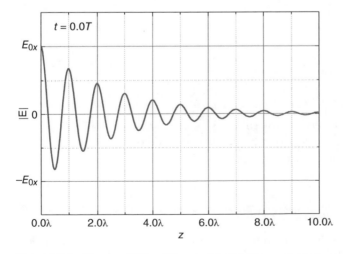

Figure 10.1 The amplitude of the electric field for an electromagnetic wave propagating in a conductor according to Eq. (10.13).

We should note that the equation for a wave propagating in the negative z-direction will have a positive rather than negative propagation constant in Eqs. (10.12) and (10.13) such that

$$\mathbf{E} = E_{0x}\mathbf{i}\exp(j\omega t)\exp(\gamma z) = E_{0x}\mathbf{i}\exp(j[\omega t + \beta z])\exp(\alpha z) \tag{10.14}$$

Although we have only worked in terms of the electric field in this section, we can of course derive a wave equation similar to Eq. (10.9) in terms of the magnetic field. In this case, the starting point is to take the curl of the Ampère–Maxwell law (Eq. (10.1d)) while also substituting the magnetic field for the magnetic flux density using $\mathbf{B} = \mu_0\mu_r\mathbf{H}$ (Eq. (4.9)) to give

$$\nabla \times \nabla \times \mathbf{B} = \mu_0\mu_r\nabla \times \left(\mathbf{J} + \frac{\partial \mathbf{D}}{\partial t}\right) \tag{10.15}$$

We can then use the same vector calculus identity (Eq. (10.3)) to re-express this as

$$\nabla(\nabla.\mathbf{B}) - \nabla^2\mathbf{B} = \nabla \times \left(\mathbf{J} + \frac{\partial \mathbf{D}}{\partial t}\right) \tag{10.16}$$

However, as we know from the Gauss law of magnetic fields that $\nabla.\mathbf{B} = 0$ (Eq. (10.1b)), the first term in this equation can be removed. Furthermore, we can use the Ohm law (Eq. (10.8)) to substitute the current density for the electric field and $\mathbf{D} = \varepsilon_0\varepsilon_r\mathbf{E}$ (Eq. (2.15)) to give

$$-\nabla^2\mathbf{B} = \mu_0\mu_r\sigma(\nabla \times \mathbf{E}) + \mu_0\mu_r\varepsilon_0\varepsilon_r\frac{\partial}{\partial t}(\nabla \times \mathbf{E}) \tag{10.17}$$

Finally, we can substitute in the Faraday law of magnetic fields (Eq. (10.1c)) so that

$$\nabla^2\mathbf{B} = \mu_0\mu_r\sigma\frac{\partial \mathbf{B}}{\partial t} + \mu_0\mu_r\varepsilon_0\varepsilon_r\frac{\partial^2 \mathbf{B}}{\partial t^2} \tag{10.18}$$

and use $\mathbf{B} = \mu_0\mu_r\mathbf{H}$ once more to produce an equivalent wave equation to Eq. (10.9) for magnetic fields

$$\nabla^2\mathbf{H} = \mu_0\mu_r\sigma\frac{\partial \mathbf{H}}{\partial t} + \mu_0\mu_r\varepsilon_0\varepsilon_r\frac{\partial^2 \mathbf{H}}{\partial t^2} \tag{10.19}$$

10.3 The Skin Effect

In Section 10.2, we saw that Eq. (10.10) was a general solution to the wave equation for conducting media (Eq. (10.9)) for a monochromatic plane wave propagating in the positive z-direction. The key difference between this and the situation in dielectric media (Eqs. (7.7) and (7.25)) is the addition of an exponential term containing the propagation constant γ which is itself given by Eq. (10.11). The propagation constant is a complex number and we said that we would look at the physical significance of the real and complex components, α and β, respectively.

We can gain this physical insight by substituting our solution (Eq. (10.10)) into the wave equation (Eq. (10.9)). It is straightforward to manipulate Eq. (10.9) to show that $\partial\mathbf{E}/\partial t = j\omega\mathbf{E}$, $\partial^2\mathbf{E}/\partial t^2 = -\omega^2\mathbf{E}$ and $\nabla^2\mathbf{E} = (\alpha + j\beta)^2\mathbf{E}$, and by substituting these into Eq. (10.9), we have that

$$(\alpha + j\beta)^2\mathbf{E} = j\omega\mu_0\mu_r\sigma\mathbf{E} - \omega^2\mu_0\mu_r\varepsilon_0\varepsilon_r\mathbf{E} \tag{10.20}$$

This allows us to equate the propagation constant to the physical properties of the conducting medium through which the wave is propagating:

$$\gamma^2 = (\alpha + j\beta)^2 = j\omega\mu_0\mu_r\sigma - \omega^2\mu_0\mu_r\varepsilon_0\varepsilon_r \tag{10.21}$$

To understand this equation we must remind ourselves that the Ampère–Maxwell law (Eq. (5.32)) contains two effective current terms, both of which can produce circulating magnetic fields. The first of these is the conduction current associated with the simple movement of charge **J**. The second is the displacement current $\partial\mathbf{D}/\partial t$ which is physically a time-varying electric flux density.

In a dielectric medium, although there is no conduction current, there is a displacement current. As $\sigma = 0$, Eq. (10.21) simplifies to

$$\gamma^2 = (\alpha + j\beta)^2 = -\omega^2\mu_0\mu_r\varepsilon_0\varepsilon_r \tag{10.22}$$

This term in the complete propagation constant equation (Eq. (10.21)) is therefore related to the displacement current caused by the electromagnetic wave passing through the medium. We should also note that taking the square root of the displacement current term alone from the propagation constant gives

$$\gamma = \alpha + j\beta = j\omega\sqrt{\mu_0\mu_r\varepsilon_0\varepsilon_r} \tag{10.23}$$

showing that in a dielectric there is no electromagnetic wave attenuation ($\alpha = 0$) and $\beta = 2\pi/\lambda = \omega\sqrt{\mu_0\mu_r\varepsilon_0\varepsilon_r}$, giving a wave velocity of $c = 1/\sqrt{\mu_0\mu_r\varepsilon_0\varepsilon_r}$ as we have proved before in dielectrics (see Section 7.3).

A conducting medium also allows a conduction current to flow as we have free charges, and such a conduction current will be induced by the electromagnetic wave as it propagates. This is the origin of the $j\omega\mu_0\mu_r\sigma$ term in the complete equation for the propagation constant (Eq. (10.21)). In practice, although there may be a small displacement current in a conductor, it is almost always insignificant compared with any conduction current. Inspection of Eq. (10.21) shows that this assumption is valid if $\sigma \gg \omega\varepsilon_0\varepsilon_r$. Relative permittivity is frequency-dependent, and so we must not confuse the ability of a conductor to completely screen static electric fields with the situation where the electric field is time varying as for an electromagnetic wave. It is very difficult to measure the polarization of a conductor under these circumstances, because of the large magnitude of the conduction current that flows, but it is reasonable to assume that this condition is met for practical conducting media. This means that we only need to consider the conduction current term in Eq. (10.21), which reduces to

$$\gamma = \alpha + j\beta = \sqrt{j\omega\mu_0\mu_r\sigma} \tag{10.24}$$

Therefore, remembering that $\sqrt{j} = (1 + j)/\sqrt{2}$, we can show that

$$\alpha = \beta = \sqrt{\frac{\omega\mu_0\mu_r\sigma}{2}} \tag{10.25}$$

The significance of this result is that it relates the attenuation term α in the propagation constant to the conductivity and permeability of the conducting medium and to the frequency of the electromagnetic wave. Specifically, Eq. (10.13) states that the electromagnetic wave decays exponentially with distance as $\exp(-\alpha z)$, and so will be attenuated by a factor of e over a distance of α^{-1}. We call this characteristic attenuation distance the *skin depth*, and it is expressed as

$$\delta = \alpha^{-1} = \sqrt{\frac{2}{\omega\mu_0\mu_r\sigma}} \tag{10.26}$$

It shows us that, for a given material, the skin depth decreases with increasing frequency. This makes intuitive sense as a greater frequency for the same amplitude results in a higher conduction current density, greater resistive (Joule heating) losses and so a more rapid transfer of energy from the electromagnetic wave into the conducting medium.

This has a very significant impact on the performance of transmission lines. By way of an example, consider the inner conductor of the coaxial cable shown in Figure 10.2. Let us imagine that we have connected an a.c. voltage source to the input end of the transmission line (we will assume that we have correctly matched the impedance of the voltage source to the cable and that there are no reflections on the cable). We know that an electromagnetic wave exists in the dielectric between the inner and outer conductors, and that this will penetrate into the conductors by a distance approximately equal to the skin depth δ. If δ is large compared with the radius of the inner conductor, then the wave will penetrate all the way to the centre and an alternating current will flow in the entire cross section of the conductor. The series resistance of the conductor per unit length (R in Figure 6.9 of the equivalent circuit of the 'lossy' transmission line) will therefore be determined by the geometry of the conductor and the resistivity of the conducting material from which it has been made. However, if δ is less than the radius, then the wave will only penetrate a small distance into the conductor, and conduction will be limited to the surface 'skin', from where the skin effect gets its name. The skin depth will now also have to be included in the calculation of the series resistance per unit length of the conductor as the effective cross section of the conductor in which current is flowing is reduced.

Figure 10.3 shows the approximate resistance per unit length of a 1 mm diameter copper wire as a function of frequency, where it is assumed that conduction is entirely limited to the skin depth. For frequencies below ~20 kHz, the resistance per unit length is a constant defined by the resistivity of the copper and the physical radius of the wire. However, above this frequency, the skin depth is less than 0.5 mm and so there is greater resistance per unit length, with an order of magnitude increase by ~7 MHz and two orders of magnitude

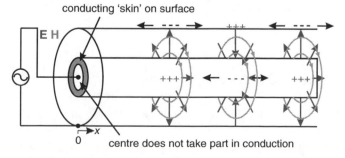

Figure 10.2 Schematic of an electromagnetic wave on a coaxial cable showing the limited cross section of the inner conductor (shaded in grey) taking part in conduction at high frequencies due to the skin effect.

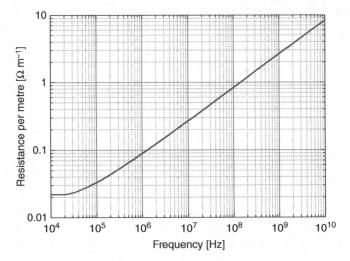

Figure 10.3 Estimation of the resistance per unit length of a copper wire of 1 mm diameter where conduction is assumed to be limited entirely to within the skin depth of the surface.

by ~700 MHz. Equation (6.51) shows that as the resistance per unit length increases, so the real part of the propagation constant also increases and the distance over which the electromagnetic wave on the transmission line is attenuated decreases.

This is why optical fibres are so important for high-speed internet connections. If a copper wire is being used to connect a home to a telephone exchange for data, then the signal received at the home from the exchange must be of sufficient amplitude. The longer the distance between the two, the lower the frequency that can be used for the signal (which limits the data rate) so that the effective resistance per unit length of the wire is low. The shorter the length of the wire, the greater the resistance per unit length that can be tolerated and so the greater the frequency that can be used and the higher the data rate. Optical fibres do not suffer from such a problem, and so offer much higher data rates than wire-based connections, but installing optical fibres to the home is expensive. Therefore a compromise is 'fibre to the cabinet' where optical fibre is used to connect telephone exchanges to street-based cabinets and then existing copper wires are used to connect from the cabinet to a home. This means that the length of the copper wire is significantly reduced, and data rates can be greatly increased.

10.4 Intrinsic Impedance of Conducting Media

When we studied the properties of electromagnetic waves in dielectric media in Section 7.5, we found that the Maxwell equations predicted that there would be a constant relating the magnitude of the electric and magnetic field components in the wave which is dependent on the permittivity and permeability of the medium. For a plane polarized wave where the electric field was in the x-direction, the magnetic field was in the y-direction and the direction of propagation was along the z-axis, the intrinsic impedance was

$$\eta = \frac{E_{0x}}{H_{0y}} = \sqrt{\frac{\varepsilon_0 \varepsilon_r}{\mu_0 \mu_r}} \tag{10.27}$$

Significantly, the electric and magnetic field components were in phase with each other, and so η was a real number.

Let us now consider the situation in conducting media by again using the monochromatic plane electromagnetic wave as our example, for which we already have an expression for the electric field in Eq. (10.13) where the electric field is in the x-direction. To determine the corresponding equation for the magnetic field (which will have to also be a valid solution to Eq. (10.19)) we need to substitute Eq. (10.13) into the Faraday law (Eq. (10.1c))

$$\nabla \times \mathbf{E} = \begin{vmatrix} \mathbf{i} & \mathbf{j} & \mathbf{k} \\ \partial/\partial x & \partial/\partial y & \partial/\partial z \\ E_{0x}\exp(j[\omega t - \beta z])\exp(-\alpha z) & 0 & 0 \end{vmatrix} = -\frac{\partial \mathbf{B}}{\partial t}$$

$$-\frac{\partial \mathbf{B}}{\partial t} = \mathbf{j}\frac{\partial}{\partial z}[E_{0x}\exp(j[\omega t - \beta z])\exp(-\alpha z)] \tag{10.28}$$

and resolving the differentiation on the right-hand side of this expression gives

$$\frac{\partial \mathbf{B}}{\partial t} = \mathbf{j}(\alpha + j\beta)E_{0x}\exp(j[\omega t - \beta z])\exp(-\alpha z) \tag{10.29}$$

Finally, we can integrate both sides of this equation with respect to time and assume that we are dealing with an ideal linear magnetic material where $\mathbf{B} = \mu_0\mu_r\mathbf{H}$ (Eq. (4.9)) to give

$$H = H_{0y}\mathbf{j}\exp(j[\omega t - \beta z])\exp(-\alpha z) \tag{10.30a}$$

where

$$H_{0y} = \frac{\alpha + j\beta}{j\omega\mu_0\mu_r}E_{0x} \tag{10.30b}$$

As for the dielectric case, this result shows that the magnetic field in the electromagnetic wave is perpendicular both to the electric field and to the direction of propagation of the wave.

We also know that the propagation constant is related to the properties of the conducting medium according to Eq. (10.21), and so we can substitute this into Eq. (10.30b):

$$H_{0y} = \sqrt{\frac{\sigma + j\omega\varepsilon_0\varepsilon_r}{j\omega\mu_0\mu_r}}E_{0x} \tag{10.31}$$

We define the intrinsic impedance η of a medium as the ratio of the electric and magnetic field components in an electromagnetic wave passing through the medium, so from Eq. (10.31) we have

$$\eta = \frac{E_{0x}}{H_{0y}} = \sqrt{\frac{j\omega\mu_0\mu_r}{\sigma + j\omega\varepsilon_0\varepsilon_r}} \tag{10.32}$$

This expression for the intrinsic impedance is completely general, applying to both dielectric and conducting media. In the case of the former, we know that $\sigma = 0$, and so this reduces back to Eq. (10.27). However, using the same assumption as in Section 10.4 that $\sigma \gg \omega\varepsilon_0\varepsilon_r$ for any reasonable conducting medium, Eq. (10.32) simplifies to

$$\eta = \sqrt{\frac{j\omega\mu_0\mu_r}{\sigma}} = (1+j)\sqrt{\frac{\omega\mu_0\mu_r}{2\sigma}} \qquad\qquad (10.33)$$

Whereas the intrinsic impedance in a dielectric medium was a real number, it is always a complex number in a conducting medium, and the magnitudes of the real and imaginary components are always equal. When we consider the impact of this on the equations for the electric and magnetic fields in the electromagnetic wave (Eqs. (10.13) and (10.30a), respectively), it is telling us that there is a phase difference between the two, with the electric field leading the magnetic field by $\pi/4$, as illustrated in Figure 10.4. The physical origin of this phase difference is the additional presence of the conduction current in the conducting medium which will flow as a result of the electromagnetic wave, compared with the dielectric medium where only a displacement current is present.

The second key area of difference between conducting media and dielectric media which emerges from Eq. (10.32) is in the magnitude of the intrinsic impedance. Most dielectrics are not magnetic, and so $\mu_r \approx 1$. Therefore, given that the relative permittivity of dielectrics is greater than unity, we know from Eq. (10.27) that the intrinsic impedance must be greater than that for a vacuum, which is 377 Ω. This means that the magnitude of the electric field in an electromagnetic wave in a dielectric is numerically greater than the magnitude of the magnetic field. Also, the relative permittivity in a dielectric is only weakly dependent on frequency, and therefore the intrinsic impedance does not have a strong frequency dependence either. However, Eq. (10.33) shows that the intrinsic impedance of conductors is frequency-dependent. Figure 10.5 shows the frequency-dependent magnitude of the intrinsic impedance for copper. As the conductivity of the medium is in the denominator of Eq. (10.33), the intrinsic impedance is significantly less than unity over a wide range of frequencies for many conductors. This means that the magnetic field component is dominant in the wave.

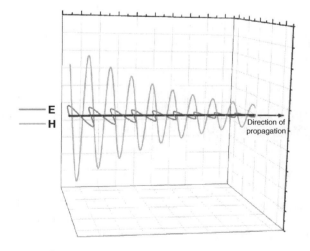

Figure 10.4 Illustration of an electromagnetic wave propagating through a conducting medium where the electric field leads the magnetic field by $\pi/4$.

E

H

Direction of propagation

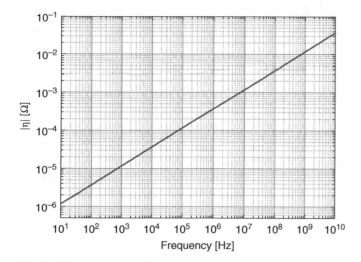

Figure 10.5 Estimation of magnitude of the intrinsic impedance of copper as a function of frequency using Eq. (10.33).

10.5 Electromagnetic Waves at Conducting Interfaces

In Chapter 9, we looked at how an electromagnetic wave behaves when it meets an interface between two different dielectrics. We saw that such an interface can lead to either or both wave reflection and transmission with refraction. We derived the Fresnel equations in Section 9.4 which quantify the proportion of the wave that is reflected and transmitted. In doing so, our starting point was to use the conservation of fields and flux densities at interfaces which are collectively described in Eq. (9.11). However, in deriving these conservation laws, we assumed that there were no free charges at the interface between the two materials as we considered both to be dielectrics, where the only charge present would be due to a surface polarization which is not free (see Section 9.2). A full analysis of the interaction of electromagnetic waves at conducting interfaces is therefore difficult. However, in practice the most common scenario that we have to consider is where an electromagnetic wave that is propagating in a dielectric medium is normally incident on a conducting surface, as previously shown in Figure 9.7. In this situation, we can simply use impedances of the two materials to determine the proportion of the wave that is transmitted into the conducting medium. We just need to remember that the intrinsic impedance of the conductor is a complex number, whereas it was purely real for the case of two dielectrics.

Let us consider the proportion of the incident wave that is transmitted. By analogy with our consideration of wave propagation between transmission lines in Section 6.7, we can write an equivalent to Eq. (6.81) for the proportion of the incident electric field E_i that is transmitted into the conductor as an electric field E_t:

$$\frac{E_t}{E_i} = \frac{2\eta_2}{\eta_2 + \eta_1} \tag{10.34}$$

where η_1 is the intrinsic impedance of the dielectric, which will usually be a real number greater than or equal to 377 Ω (the intrinsic impedance of air). We have seen in Section 10.4 that, for all reasonable conductors, the conductor's intrinsic impedance (which is η_2 in Eq. (10.34)) is a complex number with a phase angle of $\pi/4$ and a magnitude that is orders of magnitude smaller than that of a dielectric. Hence, Eq. (10.34) approximates to

$$\frac{E_t}{E_i} \approx \frac{2\eta_2}{\eta_1} \tag{10.35}$$

The proportion of the electric component of the electromagnetic wave that is transmitted into the metal is therefore very small. For example, for a 2.45 GHz Wi-Fi™ signal incident on a copper surface, only about 0.001% of the electric field will be transmitted. Also, as η_1 is real, E_t/E_i will have the same phase as just η_2, meaning that the transmitted electric field will lead the incident field by $\pi/4$.

We can now also work out the proportion of the incident magnetic field H_i that is transmitted into the conductor as a magnetic field H_t using Eq. (10.34) and remembering that in the dielectric medium $E_i/H_i = \eta_1$ while in the conducting medium $E_t/H_t = \eta_2$. Therefore, substitution of the electric field terms in Eq. (10.34) gives

$$\frac{H_t}{H_i} = \frac{2\eta_1}{\eta_2 + \eta_1} \tag{10.36}$$

This time, the difference in magnitude between the two intrinsic impedances means that this approximates to

$$\frac{H_t}{H_i} \approx 2 \tag{10.37}$$

The magnetic field component is therefore greater in the metal than in the dielectric, and the two are in phase. Also, given that the electric and magnetic fields in the incident wave are in phase, this result is consistent with the electric field leading the magnetic field in a conductor, as we found previously in Section 10.4.

In terms of the reflected wave, as both the incident and reflected waves are in the same dielectric medium, the reflection coefficients for both the electric and magnetic fields are the same and, by analogy with Eq. (6.68), are given by

$$\frac{E_r}{E_i} = \frac{H_r}{H_i} = \frac{\eta_2 - \eta_1}{\eta_2 + \eta_1} \tag{10.38}$$

We can again use the large difference in magnitude between the two intrinsic impedances to approximate this to

$$\frac{E_r}{E_i} = \frac{H_r}{H_i} = -1 \tag{10.39}$$

Therefore, the wave is reflected in antiphase with the incident wave, resulting in a node at the reflection point.

11

Waveguides

11.1 Introduction

In Chapters 7–10 of this book, we have seen how the Maxwell equations allow us to understand how we can generate electromagnetic waves in dielectric media using antennas, how they then propagate through dielectrics, and how they behave at interfaces with either other dielectrics or conducting media. Although particular antenna designs afford some preferred directionality to the direction of propagation of the electromagnetic wave, this is not a 'point-to-point' means of transmission of either data or energy, as was the case for the transmission lines which we studied in Chapter 6. Implicit in our analysis of transmission lines was the assumption that the wavelength in the direction of propagation was long compared with the lateral dimensions of the transmission line. Therefore, transmission lines have a finite range of operating frequencies depending on the dimensions of the line. Furthermore, they have the added complexity of requiring at least two conductors which run along the length of the line.

An alternative approach is to use a hollow tube of some regular cross section made from a conducting material to guide an electromagnetic wave. Such a tube is called a *waveguide,* and the basic principle is that the conductor effectively confines the electromagnetic wave inside the tube, resulting in transmission from one end to the other. Critically, these also operate over a specific range of frequencies where the wavelength is of the same order as the lateral dimensions of the waveguide. In studying these waveguides, we shall see that the boundary conditions imposed by the conducting walls of the tube have a profound effect on the propagation of the wave. In reality, as these waveguides must be machined out of metals, the range of sizes over which manufacture is practical means that they are most commonly used at microwave frequencies where the wavelength is in the range from a few millimetres to around 10 cm. The widespread use of microwaves from heating to communications means that waveguides are widely used.

The other key frequency range where we wish to be able to transmit waves from one point to another is in the infrared to visible part of the electromagnetic spectrum. In this case, we can use a long, thin length of a dielectric surrounded by a second dielectric of lower refractive index. As we saw in Section 9.3, these conditions allow for total internal reflection, and so an electromagnetic wave launched into one end of the dielectric can, under appropriate conditions, be confined within the dielectric as it propagates to the other end. Such optical fibres have enabled high-speed, high-bandwidth communications across

Electromagnetism for Engineers, First Edition. Andrew J. Flewitt.
© 2023 John Wiley & Sons Ltd. Published 2023 by John Wiley & Sons Ltd.
Companion website: www.wiley.com/go/flewitt/electromagnetism

the world. Once again, however, they only operate over a specific frequency range that is dependent on their physical cross section and the materials employed.

In this chapter we shall consider both waveguides and optical fibres, understanding how they must be engineered to accommodate specific frequencies for transmission.

11.2 Rectangular Waveguides: Geometry and Fields

We shall consider simple waveguides of uniform rectangular cross section in this section, with a general geometry shown in Figure 11.1. We will assume that the waveguide is fabricated from a good conductor, is filled with a simple dielectric with a relative permittivity ε_r and a relative permeability μ_r, and that it has cross-sectional dimensions of X and Y which point along Cartesian x- and y-axes, respectively. We will always assume that $X \geq Y$, that the waveguide is infinitely long in the direction of the z-axis, and that the wave is propagating in the positive z-direction.

Clearly, the presence of the waveguide acts to confine the electromagnetic wave in the x–y plane. This is in contrast to the free-space plane electromagnetic wave which we considered in Section 7.5. This wave had no component of either electric or magnetic field pointing in the (z-) direction of propagation, and so is called a *transverse electromagnetic* (TEM) wave. Furthermore, as the wave was uniform in the x–y plane, it could be described using the equations

$$\mathbf{E} = (E_{0x}\mathbf{i} + E_{0y}\mathbf{j})\exp\{j(\omega t - \beta z)\} \tag{11.1a}$$

$$\mathbf{H} = (H_{0x}\mathbf{i} + H_{0y}\mathbf{j})\exp\{j(\omega t - \beta z)\} \tag{11.1b}$$

where $\mathbf{i}$, $\mathbf{j}$ and $\mathbf{k}$ are the unit vectors and E_{0x} and E_{0y} are the components of the amplitude of the electric field in each of the x- and y-directions respectively, and similarly for the magnetic field for H_{0x} and H_{0y}.

The first effect of the confinement of the electromagnetic wave within the waveguide is that TEM waves cannot be supported. This is most straightforwardly demonstrated by considering a situation where the magnetic field is transverse to the direction of propagation. The walls of the waveguide mean that the magnetic field lines must form closed

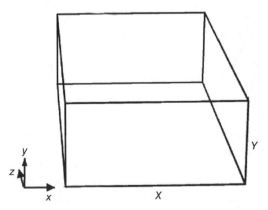

Figure 11.1 Rectangular waveguide geometry where the direction of propagation is in the positive z-direction (into the page). A short section of the infinitely long waveguide is shown.

loops within the waveguide in the *x*–*y* plane. Therefore, if we apply the Ampère–Maxwell equation in integral form (Eq. (5.34)) around any Amperian loop that is also in the *x*–*y* plane within the waveguide (i.e. following any such magnetic field line), then clearly the left-hand side of

$$\oint_C \mathbf{H.dr} = I + \int_S \frac{\partial \mathbf{D}}{\partial t}.\mathbf{dA} \tag{5.34}$$

will be non-zero. However, as the waveguide is filled with a dielectric (there is no central conductor as for a coaxial cable) then the current term on the right-hand side of this equation must be zero. The Ampère–Maxwell equation could only then be satisfied if $\int_S \frac{\partial \mathbf{D}}{\partial t}.\mathbf{dA} \neq 0$, but this would only be true if $\mathbf{D}$ has a component in the direction perpendicular to the *x*–*y* plane of the Amperian loop, which is the direction of propagation, which is not the case for a TEM wave. Therefore, we can only have propagation of waves where *either* the electric field is transverse but the magnetic field has a component in the direction of propagation *or* the magnetic field is transverse but the electric field has a component in the direction of propagation. The first of these is known as a *transverse electric* (TE) wave and the latter as a *transverse magnetic* (TM) wave.

11.3 Rectangular Waveguides: TE Modes

We must now turn our attention to finding new sets of equations for the waves in the waveguide that are similar to those for the free-space TEM waves of Eq. (11.1), but which capture the confinement in the *x*–*y* plane in addition to their TE or TM nature. We will first consider the more common TE wave, which is shown in Figure 11.2.

As the electric field is not now uniform in the *x*–*y* plane, we must adapt Eq. (11.1a) so that the E_{0x} and E_{0y} are functions of *x* and *y* as opposed to simple constants. In considering the interaction of static electric fields with metals in Section 2.6, we saw that electric fields must meet metal surfaces perpendicularly because the metal is at an equipotential. In considering time-varying fields in Chapter 10, we now know that the situation is slightly more complex than this, but it remains a very good approximation that there is no tangential component of electric field at a metal surface. Therefore, the form of E_{0x} must be such that $E_{0x} = 0$ at $y = 0$ and $y = Y$ where it tangentially meets a metal surface. This is rather like a wave on a guitar string that is clamped at either end to give a node at these points. Therefore, we would expect E_{0x} to be of the form $\sin(n\pi y/Y)$ where *n* is an integer. The form of E_{0x} in terms of *x* is perhaps less clear. One might expect it to be a maximum where it perpendicularly meets a metal surface with free surface charge, and indeed we do find that this is the case for the wave equation in free space (Eq. (7.7)) to be satisfied. It is of the form $\cos(m\pi x/X)$ where *m* is an integer. The *y*-component of the electric field is tangential with the waveguide surfaces at $x = 0$ and $x = X$, and similarly this has the form $\sin(m\pi x/X)\cos(n\pi y/Y)$. We then need to allow for the fact that each of these two components will have different amplitudes, A_{Ex} and A_{Ey} respectively, resulting in a new form of Eq. (11.1),

$$\mathbf{E} = (A_{Ex}\cos(m\pi x/X)\sin(n\pi y/Y)\mathbf{i} + A_{Ey}\sin(m\pi x/X)\cos(n\pi y/Y)\mathbf{j})\exp\{j(\omega t - \beta_g z)\}$$

$$\tag{11.2}$$

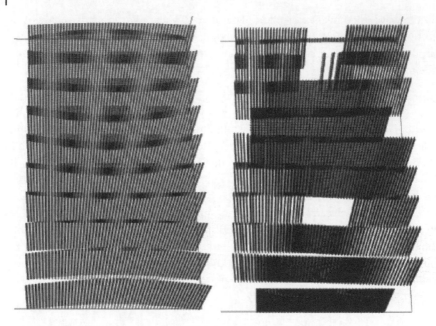

Figure 11.2 The (left) electric and (right) magnetic fields in vector form for a TE_{10} mode along a waveguide of the same geometry as in Figure 11.1. The fields are shown in layers in the $x-y$ plane and the z-axis is running from the bottom to the top of the page, so that we are effectively looking down on the waveguide from above.

where β_g is the wavenumber of the electromagnetic wave in the waveguide and is related to the wavelength in the waveguide by $\lambda_g = 2\pi/\beta_g$. We can use the Gauss law of electric fields to determine how A_{Ex} and A_{Ey} are related. As there are no free charges in the waveguide and it is filled with a simple dielectric medium, Eq. (5.7) can be rewritten as

$$\nabla.\mathbf{E} = 0 \tag{11.3}$$

We are dealing with a TE wave, so there is no component of $\mathbf{E}$ in the z-direction, and Eq. (11.3) becomes

$$\frac{\partial E_x}{\partial x} + \frac{\partial E_y}{\partial y} = 0 \tag{11.4}$$

Applying this to Eq. (11.2) (and cancelling the common $\exp\{j(\omega t - \beta_g z)\}$ term) gives

$$-\frac{A_{Ex}m\pi}{X} \sin\left(\frac{m\pi x}{X}\right)\sin\left(\frac{n\pi y}{Y}\right) - \frac{A_{Ey}n\pi}{Y} \sin\left(\frac{m\pi x}{X}\right)\sin\left(\frac{n\pi y}{Y}\right) = 0 \tag{11.5}$$

Hence,

$$A_{Ey} = -A_{Ex}\frac{mY}{nX} \tag{11.6}$$

giving us a final form of Eq. (11.2) with just a single constant A that is dependent on amplitude:

$$\mathbf{E} = A \exp\{j(\omega t - \beta_g z)\} \left[\cos\left(\frac{m\pi x}{X}\right)\sin\left(\frac{n\pi y}{Y}\right)\mathbf{i} - \frac{mY}{nX}\sin\left(\frac{m\pi x}{X}\right)\cos\left(\frac{n\pi y}{Y}\right)\mathbf{j}\right]$$
$$\tag{11.7}$$

We can now determine the magnetic field from this equation for the electric field by applying the Faraday law of magnetic fields (Eq. (5.17)) rewritten for a simple magnetic medium as

$$\nabla \times \mathbf{E} = -\mu_0 \mu_r \frac{\partial \mathbf{H}}{\partial t} \tag{11.8}$$

As the resulting equation for $\mathbf{H}$ will have an $\exp(j\omega t)$ time-dependence, the final effect of integrating the whole equation with respect to time will be the same as rewriting Eq. (11.8) as

$$\nabla \times \mathbf{E} = -\mu_0 \mu_r j\omega \mathbf{H} \tag{11.9}$$

Therefore, we must solve

$$
\begin{vmatrix}
\mathbf{i} & \mathbf{j} & \mathbf{k} \\
\partial/\partial x & \partial/\partial y & \partial/\partial z \\
A e^{j(\omega t - \beta_g z)} \cos\left(\frac{m\pi x}{X}\right)\sin\left(\frac{n\pi y}{Y}\right) & -A e^{j(\omega t - \beta_g z)}\frac{mY}{nX}\sin\left(\frac{m\pi x}{X}\right)\cos\left(\frac{n\pi y}{Y}\right)\mathbf{j} & 0
\end{vmatrix}
$$

with the result that the complementary equation for $\mathbf{H}$ is

$$
\mathbf{H} = \frac{A}{\mu_0 \mu_r}\exp\{j(\omega t - \beta_g z)\}\left[\frac{\beta_g}{\omega}\frac{mY}{nX}\sin\left(\frac{m\pi x}{X}\right)\cos\left(\frac{n\pi y}{Y}\right)\mathbf{i} + \frac{\beta_g}{\omega}\cos\left(\frac{m\pi x}{X}\right)\right.
$$
$$
\left.\sin\left(\frac{n\pi y}{Y}\right)\mathbf{j} - \frac{j}{\omega}\left(\frac{m^2\pi Y}{nX^2} + \frac{n\pi}{Y}\right)\cos\left(\frac{m\pi x}{X}\right)\cos\left(\frac{n\pi y}{Y}\right)\mathbf{k}\right] \tag{11.10}
$$

Inspection of this result shows some important features of TE waves in waveguides. Most notable is that a component of the magnetic field in the direction of propagation naturally emerges as the electric field is spatially varying in the x–y plane and hence $\partial E_y/\partial x$ and $\partial E_x/\partial y$ are both now non-zero (whereas they were zero for the unconfined plane wave in Section 7.3).

In this derivation for the form of the TE wave, we assumed that $m \neq 0$ and $n \neq 0$ in producing Eq. (11.6) for the relation between the amplitudes of the x- and y-components of the electric field. From Eq. (11.7), it is clear that we cannot have a situation where $m = n = 0$, otherwise there would be no wave ($\mathbf{E} = 0$). However, just one of these two can be zero.

For the situation where $n = 0$, Eq. (11.7) for the electric field simplifies to

$$\mathbf{E} = A\exp\{j(\omega t - \beta_g z)\}\sin\left(\frac{m\pi x}{X}\right)\mathbf{j} \tag{11.11}$$

We should note that there can be no equivalent to Eq. (11.6) as there is no x-component of the electric field. The complementary equation for $\mathbf{H}$ which is equivalent to Eq. (11.10) is then

$$\mathbf{H} = \frac{A}{\mu_0 \mu_r}\exp\{j(\omega t - \beta_g z)\}\left[-\frac{\beta_g}{\omega}\sin\left(\frac{m\pi x}{X}\right)\mathbf{i} + \frac{j}{\omega}\frac{m\pi}{X}\cos\left(\frac{m\pi x}{X}\right)\mathbf{k}\right] \tag{11.12}$$

Similarly, when $m = 0$, Eq. (11.7) simplifies to

$$\mathbf{E} = A\exp\{j(\omega t - \beta_g z)\}\sin\left(\frac{n\pi y}{Y}\right)\mathbf{i} \tag{11.13}$$

and the complementary equation for **H** is then

$$\mathbf{H} = \frac{A}{\mu_0 \mu_r} \exp\{j(\omega t - \beta_g z)\} \left[\frac{\beta_g}{\omega} \sin\left(\frac{n\pi y}{Y}\right)\mathbf{j} - \frac{j}{\omega}\frac{n\pi}{Y} \cos\left(\frac{n\pi y}{Y}\right)\mathbf{k} \right] \qquad (11.14)$$

By way of an example, Figure 11.2 shows the electric and magnetic field vectors for a short length of a transmission line with a TE_{10} mode wave (where $m = 1$ and $n = 0$). In this case, the electric field vector is always parallel to the y-axis and varies in magnitude sinusoidally along the length of the transmission line. Furthermore, it varies in magnitude across the width of the waveguide (along the x-axis), being zero in magnitude at the waveguide walls and a maximum in the centre. The magnetic field, however, circulates in the x–y plane. It has no component in the y-direction and has its minimum in the centre of the waveguide.

11.4 Rectangular Waveguides: TM Modes

Let is now consider the form of the TM waves. In this case the magnetic field has no component in the z-direction. As the lines of magnetic field must form closed loops and be contained within the waveguide, they must loop around within any cross section. Therefore, the lines of magnetic field meet the walls of the waveguide tangentially in the x–y plane. This means that $H_{0x} = 0$ at $x = 0$ and $x = X$ and also that $H_{0y} = 0$ at $y = 0$ and $y = Y$ (so that there is no perpendicular component of magnetic field at the waveguide walls). To satisfy this, the magnetic field has the form

$$\mathbf{H} = (A_{Hx} \sin(m\pi x/X)\cos(n\pi y/Y)\mathbf{i} + A_{Hy}\cos(m\pi x/X)\sin(n\pi y/Y)\mathbf{j})\exp\{j(\omega t - \beta_g z)\}$$
$$(11.15)$$

This time, we can use the Gauss law of magnetic fields (Eq. (5.7)) to determine how A_{Hx} and A_{Hy} are related, which for simple magnetic materials can be written as

$$\nabla.\mathbf{H} = 0 \qquad (11.16)$$

As we are dealing with a TM wave, there is no component of **H** in the z-direction, and so Eq. (11.16) becomes

$$\frac{\partial H_x}{\partial x} + \frac{\partial H_y}{\partial y} = 0 \qquad (11.17)$$

Applying this to Eq. (11.15) (and again cancelling the common $\exp\{j(\omega t - \beta_g z)\}$ term) gives a very similar expression to Eq. (11.6) for the TE wave,

$$A_{Hy} = -A_{Hx}\frac{mY}{nX} \qquad (11.18)$$

This gives us a form of Eq. (11.15) with just a single constant A that is dependent on amplitude

$$\mathbf{H} = A \exp\{j(\omega t - \beta_g z)\} \left[\sin\left(\frac{m\pi x}{X}\right)\cos\left(\frac{n\pi y}{Y}\right)\mathbf{i} - \frac{mY}{nX}\cos\left(\frac{m\pi x}{X}\right)\sin\left(\frac{n\pi y}{Y}\right)\mathbf{j} \right]$$
$$(11.19)$$

This time, we can use the Ampère–Maxwell law (Eq. (5.32)) to find the complementary equation for the electric field. As there is just a simple dielectric filling the waveguide, there can be no current flow and the Ampère–Maxwell law becomes

$$\nabla \times \mathbf{H} = \varepsilon_0 \varepsilon_r \frac{\partial \mathbf{E}}{\partial t} \tag{11.20}$$

With the same methodology that we used to find the equation for $\mathbf{H}$ from $\mathbf{E}$ for the TE waves, we find

$$\mathbf{E} = \frac{A}{\varepsilon_0 \varepsilon_r} \exp\{j(\omega t - \beta_g z)\} \left[\frac{\beta_g}{\omega} \frac{mY}{nX} \cos\left(\frac{m\pi x}{X}\right) \sin\left(\frac{n\pi y}{Y}\right) \mathbf{i} - \frac{\beta_g}{\omega} \sin\left(\frac{m\pi x}{X}\right) \right.$$
$$\left. \times \cos\left(\frac{n\pi y}{Y}\right) \mathbf{j} + \frac{j}{\omega} \left(\frac{m^2 \pi Y}{nX^2} + \frac{n\pi}{Y} \right) \sin\left(\frac{m\pi x}{X}\right) \sin\left(\frac{n\pi y}{Y}\right) \mathbf{k} \right] \tag{11.21}$$

In a similar fashion to the analysis of the TE wave, we see that the spatial variation of the magnetic field in the x–y plane means that $\partial H_y/\partial x$ and $\partial H_x/\partial y$ are non-zero, leading to a component of the electric field (this time) in the direction of propagation.

Comparing the equations for the TE wave (Eqs. (11.7) and (11.10)) with those for the TM wave (Eqs. (11.19) and (11.21)) reveals that the two are mathematically very similar. There is always an integer number of half cycles in the variation of the field that is transverse across the waveguide, but the sine and cosine terms are swapped because of the different boundary conditions for the electric and magnetic fields (Eqs. (11.7) and (11.10)). This also results in a swapping of the sine and cosine terms for the field that has a component in the direction of propagation (Eqs. (11.19) and (11.21)), but the consequence of this for the TM wave is more profound as the component of $\mathbf{E}$ in the (z-) direction of propagation is dependent on $\sin(m\pi x/X)\sin(n\pi y/Y)$. This means that if either m or n is zero, then there is no component of electric field in this direction; this would result in a TEM wave which, as we have seen, is not physically allowed. Hence, the lowest possible mode for TM waves is $m = n = 1$, and Eqs. (11.19) and (11.21) describe all possible TM waves (there are no special cases as for TE waves).

11.5 Rectangular Waveguides: Propagation of Modes

We now have a complete set of equations describing the electromagnetic waves, both TE and TM, that can propagate in a rectangular waveguide. Inspection of these equations shows that they all have a similar form,

$$\mathbf{F} = \mathbf{f}\{x, y\} \exp[j(\omega t - \beta_g z)] \tag{11.22}$$

where $\mathbf{F}$ is either the electric or magnetic field. $\mathbf{f}\{x, y\}$ is a function describing the spatial variation of the field across the waveguide that is always only dependent on x and y; although it may define the magnitude of the component in the z-direction, it is never dependent on the value of z. Instead, the variation in the field along the length of the waveguide is entirely described by the exponential part of Eq. (11.22). In this regard, the propagation constant of the wave within the waveguide β_g is important. We should remember that the propagation constant is the spatial equivalent of angular frequency (Eq. (6.20)) and hence is related to wavelength in the waveguide λ_g by

$$\beta_g = \frac{2\pi}{\lambda_g} \tag{11.23}$$

In using the term β_g for the propagation constant in the waveguide, we are explicitly allowing the possibility that the speed of propagation in a waveguide filled with a particular dielectric may be different than a wave of the same frequency propagating through the same dielectric but outside the confines of the waveguide (i.e. the presence of the waveguide alters the speed of propagation).

To explore this further, we need to return to the wave equation for an electromagnetic wave propagating in a dielectric, which we derived in Section 7.3 from the Maxwell equations allowing for the medium being a non-conducting dielectric. As this condition is still true inside the waveguide, Eqs. (7.7) and (7.13) for the electric and magnetic fields remain valid. Both have the form

$$\nabla^2 \mathbf{F} = \varepsilon_0 \varepsilon_r \mu_0 \mu_r \frac{\partial^2 \mathbf{F}}{\partial t^2} \tag{11.24}$$

which in Cartesian coordinates becomes

$$\frac{\partial^2 \mathbf{F}}{\partial x^2} + \frac{\partial^2 \mathbf{F}}{\partial y^2} + \frac{\partial^2 \mathbf{F}}{\partial z^2} = \varepsilon_0 \varepsilon_r \mu_0 \mu_r \frac{\partial^2 \mathbf{F}}{\partial t^2} \tag{11.25}$$

As the only time-dependence and z-dependence of $\mathbf{F}$ is in the exponential term (Eq. (11.22)), we can replace $\partial^2 \mathbf{F}/\partial t^2$ with $-\omega^2 \mathbf{F}$ and $\partial^2 \mathbf{F}/\partial z^2$ with $-\beta_g^2 \mathbf{F}$, giving

$$\frac{\partial^2 \mathbf{F}}{\partial x^2} + \frac{\partial^2 \mathbf{F}}{\partial y^2} - \beta_g^2 \mathbf{F} = -\varepsilon_0 \varepsilon_r \mu_0 \mu_r \omega^2 \mathbf{F} \tag{11.26}$$

We know from Eq. (7.9) that $1/\sqrt{\varepsilon_0 \varepsilon_r \mu_0 \mu_r}$ is the velocity of the electromagnetic wave in the dielectric medium c. Therefore, using Eq. (6.23),

$$\varepsilon_0 \varepsilon_r \mu_0 \mu_r \omega^2 = \frac{\omega^2}{c^2} = \beta^2 \tag{11.27}$$

where β is the propagation constant that an electromagnetic wave of the same frequency would have in the same dielectric but outside of the waveguide. Also, the exponential part of $\mathbf{F}$ has no x- or y-dependence, so by eliminating the exponential as a common term from Eq. (11.26) and substituting Eq. (11.27), we are left with

$$\frac{\partial^2 \mathbf{f}}{\partial x^2} + \frac{\partial^2 \mathbf{f}}{\partial y^2} - \beta_g^2 \mathbf{f} = -\beta^2 \mathbf{f} \tag{11.27}$$

If we had the simple situation that we had in Section 7.5 where we were considering plane electromagnetic waves, there would be no variation in $\mathbf{f}$ in the x–y plane perpendicular to the direction of propagation, and the first two terms in Eq. (11.27) would be zero, resulting in $\beta_g = \beta$. However, in the waveguide $\mathbf{f}$ is a function of x and y. Moreover, inspection of the expression for the electric and magnetic field in any of the TE or TM waves (Eqs. (11.7), (11.10), (11.11), (11.12), (11.13), (11.14), (11.19), and (11.21)) reveals that $\mathbf{f}$ is always of the form

$$\mathbf{f} = f_x \sin/\cos\left(\frac{m\pi x}{X}\right) \sin/\cos\left(\frac{n\pi y}{Y}\right) \mathbf{i} + f_y \sin/\cos\left(\frac{m\pi x}{X}\right)$$
$$\sin/\cos\left(\frac{n\pi y}{Y}\right) \mathbf{j} + f_z \sin/\cos\left(\frac{m\pi x}{X}\right) \sin/\cos\left(\frac{n\pi y}{Y}\right) \mathbf{k} \tag{11.28}$$

where f_x, f_y and f_z are simple constants and sin/cos represents either a sine or cosine function. As $\partial^2(\sin[\alpha x])/\partial x^2 = -\alpha^2 \sin[\alpha x]$ and $\partial^2(\cos[\alpha x])/\partial x^2 = -\alpha^2 \cos[\alpha x]$ (i.e. the second differential of either trigonometric function returns the original function multiplied by minus a constant squared) we can evaluate the second differential in Eq. (11.27) for all of the possible TE and TM waves, giving

$$-\left(\frac{m\pi}{X}\right)^2 \mathbf{f} - \left(\frac{n\pi}{Y}\right)^2 \mathbf{f} - \beta_g^2 \mathbf{f} = -\beta^2 \mathbf{f} \tag{11.29}$$

which rearranges to

$$\beta_g^2 = \beta^2 - \left(\frac{m\pi}{X}\right)^2 - \left(\frac{n\pi}{Y}\right)^2 \tag{11.30}$$

We therefore have an equation for the propagation constant of any TE or TM wave in a waveguide in terms of the propagation constant of the wave of the same frequency outside of the waveguide, the numerical mode of the wave expressed in terms of m and n, and the dimensions of the waveguide. Hence, the wavelength of the wave is dependent on the mode as well as the frequency and the dielectric medium. This is perhaps intuitive as we are setting up a periodicity in the electric and magnetic fields across the waveguide, but Eq. (11.30) makes this explicit. Furthermore, as m and n are positive integers, it must be the case that $\beta_g < \beta$, or, in other words, the wavelength inside the waveguide is greater than that outside the waveguide. As the frequency is the same, this means that the velocity of the wave in the waveguide is increased. This might suggest that we can break Einstein's special theory of relativity, but we should remember that this calculation gives the phase velocity, whereas the velocity with which energy or information is transmitted is the group velocity, which will be less than the speed of light in a vacuum inside the waveguide (see Section 7.4).

Let us therefore image a situation where we have an antenna in free space which is producing an electromagnetic wave with a propagation constant β and we are going to see if it will propagate along a waveguide in a particular mode given by m and n. It is clear from Eq. (11.30) that

$$\beta^2 > \left(\frac{m\pi}{X}\right)^2 + \left(\frac{n\pi}{Y}\right)^2 \tag{11.31}$$

if $\beta_g^2 > 0$. This must be the case if β_g is to be a real number; if it was imaginary then the exponential term in Eq. (11.22) would become a simple exponential decay with z and the wave would be rapidly attenuated (i.e. it would not propagate). Given that the propagation constant is related to angular frequency by $\beta = \omega/c$, this means that there is a cut-off frequency ω_{co} for any given mode, which is the minimum frequency for propagation to occur:

$$\omega_{co} = \sqrt{\frac{1}{\varepsilon_0 \varepsilon_r \mu_0 \mu_r} \left[\left(\frac{m\pi}{X}\right)^2 + \left(\frac{n\pi}{Y}\right)^2\right]} \tag{11.32}$$

When we defined the geometry of the waveguide at the start of this section, we said that we would always assume that $X \geq Y$, and so the TE_{10} mode ($m = 1$, $n = 0$) has the lowest cut-off frequency, which is

$$\omega_{co10} = \sqrt{\frac{\pi}{\varepsilon_0 \varepsilon_r \mu_0 \mu_r X}} \tag{11.33}$$

Table 11.1 The lowest eight cut-off frequencies for modes in a rectangular waveguide with dimensions of $X = 72.136$ mm and $Y = 34.036$ mm (type WR284) calculated using Eq. (11.33).

Mode	Cut-off frequency (GHz)
TE_{10}	2.078
TE_{20}	4.156
TE_{01}	4.404
TE_{11} and TM_{11}	4.870
TE_{21} and TM_{21}	6.055
TE_{30}	6.234
TE_{31} and TM_{31}	7.633

In practice, it is preferable for a wave to propagate in a single 'dominant' mode in a waveguide. For this reason, the TE_{10} mode is almost exclusively used. This is because, depending on dimensions, there will be a range of frequencies greater than ω_{co10} over which only the TE_{10} mode can propagate as the frequency is below the cut-off frequency of the next highest mode.

Launching a wave of this frequency into the waveguide within this range will result in a TE_{10} mode only. For example, Table 11.1 shows the cut off frequencies for the lowest eight modes in a rectangular waveguide with dimensions of $X = 72.136$ mm and $Y = 34.036$ mm (commonly called a WR284 or R32 waveguide). In this case, the TE_{10} mode has a cut-off frequency of 2.078 GHz, and so any wave with a frequency less than this will not propagate through the waveguide. The next lowest cut-off frequency is 4.156 GHz, which is associated with the TE_{20} mode. Therefore, waves with frequencies in the range between 2.078 and 4.156 GHz will only propagate in the TE_{10} mode; this is the frequency range over which this waveguide is really designed to be used.

11.6 Rectangular Waveguides: Power, Impedance and Attenuation

We now have a full set of equations in terms of the electric and magnetic fields for the different electromagnetic wave modes that can exist in a waveguide and a way of determining which modes may exist (if any) at a given frequency. So far, however, we have not looked at how much power is being propagated along the waveguide. Instead, we have always had a constant representing the amplitude of the waves in our various equations for the electric and magnetic fields, being careful in every case to ensure that we only define the amplitude for one field and then calculate the amplitude of the other from that. Understanding the equations for power transmission would then allow us to calculate what amplitude we would need to achieve a particular power.

In Section 7.7, we saw that the Poynting vector, which is given by

$$\mathbf{N} = \mathbf{E} \times \mathbf{H} \tag{7.45}$$

expresses the instantaneous power per unit area of an electromagnetic wave in vector form, so we know both the instantaneous magnitude and the direction. We solely had TEM waves in free space where the electric and magnetic fields only had components which were perpendicular to each other and to the direction of propagation. In the waveguide, this is not the case as either the magnetic (in TE waves) or electric (in TM waves) field has a component in the direction of propagation, but this component will not lead to a propagation of power along the length of the waveguide. Only the z- component of the Poynting vector N_z contributes to a power flow along the waveguide. The instantaneous power along the waveguide at any point in a cross section is therefore

$$N_z = E_x H_y - E_y H_x \tag{11.34}$$

To calculate the total power flow P along the waveguide, we then need to average N_z over one wave cycle (i.e. from $\omega t = 0$ to $\omega t = 2\pi$) and to integrate over a cross section,

$$P = \frac{1}{2\pi} \int_0^Y \int_0^X \int_0^{2\pi} N_x \, d(\omega t) \, dx \, dy \tag{11.35}$$

By way of an example, let us consider the most common TE_{10} mode whose electric and magnetic fields can be expressed from Eqs. (11.11) and (11.12) as

$$\mathbf{E} = A \exp\{j(\omega t - \beta_g z)\} \sin\left(\frac{\pi x}{X}\right) \mathbf{j} \tag{11.36a}$$

$$\mathbf{H} = \frac{A}{\mu_0 \mu_r} \exp\{j(\omega t - \beta_g z)\} \left[-\frac{\beta_g}{\omega} \sin\left(\frac{\pi x}{X}\right) \mathbf{i} + \frac{j}{\omega} \frac{\pi}{X} \cos\left(\frac{\pi x}{X}\right) \mathbf{k}\right] \tag{11.36b}$$

As there is no x-component to the electric field, Eq. (11.34) simplifies to

$$N_z = -E_y H_x$$

$$= A \exp\{j(\omega t - \beta_g z)\} \sin\left(\frac{\pi x}{X}\right) \frac{A}{\mu_0 \mu_r} \exp\{j(\omega t - \beta_g z)\} \frac{\beta_g}{\omega} \sin\left(\frac{\pi x}{X}\right) \tag{11.37}$$

For the purposes of determining the average power, it is easier to express the complex exponential in terms of just the real part, using the de Moivre theorem to give

$$N_z = \frac{A^2}{\mu_0 \mu_r} \frac{\beta_g}{\omega} \sin^2\left(\frac{\pi x}{X}\right) \cos^2(\omega t - \beta_g z) \tag{11.39}$$

The average over time of $\cos^2(\omega t - \beta_g z)$ is just $1/2$, and so Eq. (11.35) for the power transmitted along the waveguide is then simply

$$P = \frac{A^2 \beta_g}{2\mu_0 \mu_r \omega} \int_0^Y \int_0^X \sin^2\left(\frac{\pi x}{x}\right) dx \, dy$$

$$= \frac{A^2 \beta_g XY}{4\mu_0 \mu_r \omega} \tag{11.40}$$

Therefore, knowing the amplitude of the TE_{10} wave allows us to calculate the power transmitted, and vice versa.

When considering other scenarios for electromagnetic wave propagation, whether in free space, dielectrics, conductors or even transmission lines, we have been able to define either an intrinsic or characteristic impedance, and this has allowed us to simplify calculations. Being able to define a characteristic impedance for a waveguide also looks attractive as it should allow us to understand how easily a wave can be coupled from a source into a waveguide, or from a waveguide into some load or via an antenna into free space. Trying to adapt the approach from free-space electromagnetic waves by using the ratio of the electric and magnetic fields to define an impedance might seem attractive at first sight, but the forms of the equations for the electric and magnetic fields are not the same across any x–y plane. This is the case even for the simple TE_{10} mode that has a spatial cosine term in the expression for the magnetic field (Eq. (11.36b)) which does not appear in that for the electric field (Eq. (11.36a)). Therefore, attributing a characteristic impedance to the waveguide is not going to have the same clear physical meaning that we have seen in other scenarios.

One approach is to consider the purpose of attributing a characteristic impedance to a waveguide as being to assist in understanding power transfer. If we do this, then a sensible definition of waveguide impedance Z_g would be to draw on our known relation between power and voltage:

$$Z_g = \frac{P}{|V|^2} \tag{11.41}$$

where $|V|$ is the magnitude of the potential difference between opposite sides of the waveguide. For example, in the case of the TE_{10} mode, we can calculate $|V|$ by integrating the electric field expressed in Eq. (11.36a) along the x-direction, but taking $\exp\{j(\omega t - \beta_g z)\} = 1$ to give the magnitude

$$|V| = \int_0^X A \sin\left(\frac{\pi x}{X}\right) dx = \frac{2XA}{\pi} \tag{11.42}$$

Hence, substituting this and our expression for power in Eq. (11.40) into Eq. (11.41) gives

$$Z_g = \frac{16X\mu_0\mu_r\omega}{Y\beta_g} \tag{11.43}$$

This expression is clearly specific both for a given wave mode (in this case TE_{10}) and frequency, and for the case of the WR284 waveguide in Table 11.1 operating at a common microwave frequency of 2.45 GHz is numerically equal to 24.1 kΩ. In considering the form of Eq. (11.43), it might at first appear as though it is only the ratio of the cross-sectional dimensions of the waveguide that is important in determining the waveguide impedance, but we should remember that, from Eq. (11.30), the dimensions also affect β_g, and specifically the dimension X for the TE_{10} mode. Therefore, if we need to match the impedance of a waveguide to that of a source or load, one means of doing so is to gradually change the dimensions towards the end of the waveguide. Alternatively, we need to change the form of the electromagnetic wave inside the waveguide towards the end but without changing its overall dimensions. This can be achieved by placing a conductor inside the waveguide, such as a post.

Finally, we should remember that some energy will always be dissipated inside a waveguide: either by inducing a current in the walls of the waveguide which will result in power dissipation by Joule heating, or by losses directly into a dielectric medium itself. In either circumstance, the result is that the wave will exponentially decay with distance, and this is achieved mathematically by adding an exp($-\alpha z$) term to all of the equations for the electric and magnetic fields where α is the attenuation coefficient, which is equal to the reciprocal of the distance over which the amplitude decays by a factor of e.

11.7 Optical Fibres

Although we have exclusively considered rectangular waveguides in this chapter so far, it is possible to construct waveguides with other cross sections. The most ubiquitous waveguide, the optical fibre, is most commonly produced with a circular cross section. Optical fibres are the backbone for transmission of data at high rates as, unlike with transmission lines, electromagnetic waves of very high frequency can be used. Today, the world is encircled with optical fibres running both over land and under the sea, allowing the almost instantaneous communication that we now take for granted.

The optical fibre consists of a central dielectric of circular cross section and refractive index n_1 that is surrounded by a sheath of a second dielectric with a refractive index n_2. The entire fibre is usually coated in a mechanical protection layer too. We need light to be confined within the optical fibre by total internal reflection, and therefore we need $n_1 > n_2$, as shown in Figure 11.3. Therefore, from our understanding of the Snell law (see Section 9.3) we know that for this to occur we require the angle of incidence of the wave in the central dielectric with the sheath dielectric to satisfy

$$\theta_i \geq \sin^{-1}\left(\frac{n_2}{n_1}\right) \tag{9.16}$$

If we consider this in terms of the angle that an incoming wave makes with the axis of the fibre θ, as shown in Figure 11.3, then this condition can be rewritten as

$$n_1 \sin \theta \leq \sqrt{n_1^2 - n_2^2} \tag{11.44}$$

This quantity is known as the *numerical aperture* of the fibre. It is frequently used in optics as it defines the angle over which light will enter the system in such a way that it will propagate.

Just as with the rectangular metal waveguide that we have considered previously, we should solve the wave equation with appropriate boundary conditions for the interface between the core and sheath dielectric in order to understand the conditions under which light will propagate along a fibre. This is best done using cylindrical polar coordinates, but even then it is a complex derivation with a number of modes emerging as being possible,

Figure 11.3 Schematic diagram of an optical fibre showing the key angles for propagation.

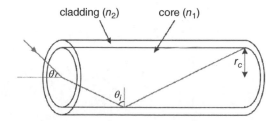

rather as for the rectangular waveguide, and with the radial term usually being in the form of a Bessel function (the angular term will be periodic over 2π). These modes have both an electric and magnetic field component pointing along the axis of the fibre, but one will be dominant in this direction. Modes where the electric field dominates are called *EH modes* and those where the magnetic field dominates are called *HE modes*. The larger the radius of the core dielectric r_c relative to the wavelength of the electromagnetic wave, the more modes become possible.

For this reason, in practice it is the so-called *V-parameter* of the fibre that is important. This is given by

$$V = \beta r_c N_{NA} \tag{11.45}$$

where β is the wavenumber ($2\pi/\lambda$) and N_{NA} is the numerical aperture. The minimum V for one single mode to be supported in a fibre is 2.405. Such *single-mode* fibres have become the standard for long-distance communication networks. The challenge is that long-distance fibres also require that the core dielectric has minimal optical absorption at the wavelength being used, which then limits the range of materials that can be used. Silicon dioxide is therefore widely used (sometimes with the incorporation of germanium to modulate the refractive index) as this has a low absorption at key near infrared wavelengths around ~1.3 and ~1.5 μm. As the numerical aperture and wavelength are therefore defined, this means that the radius of the fibre core must be only a few micrometres.

Historically, single-mode fibres were challenging to manufacture and so larger-radius *multi-mode* fibres were employed. In this case, the number of modes that can be accommodated is approximately $V^2/2$. Multi-mode fibres have the inherent problem that, just as for waves on a rectangular waveguide, the velocity of the wave depends on the mode. Therefore, the pulses of light that are used to encode data, and which will exist in several modes in a multi-mode fibre, will suffer from dispersion (see Section 7.4) so that, over longer distances, the pulses become smeared out and eventually undetectable. Today, multi-mode fibres are still sometimes used for local networks.

12

Three-Phase Electrical Power

12.1 Introduction

A constant theme throughout this book has been the impact that electromagnetism has on our everyday lives through the work of scientists and engineers in transforming fundamental concepts into real applications. Although we have focused very heavily on communications up to now, in this last chapter we turn our attention to the generation, transmission and use of three-phase electrical power. The development of electrical power distribution systems which decouple where electricity is generated from where it is consumed is one of the single most important steps in our technological development. This has not only allowed greater efficiency to be achieved through power generation at scale, but more recently it is the enabler of a new, carbon-zero future where renewable energy supplies can be harnessed where they are available (e.g. wind turbines at sea or tidal power plants in large river estuaries) and the energy consumed wherever it is needed.

Although some high-power d.c. links are used (most commonly connecting the power grids of different countries together and frequently under the sea), a.c. systems dominate the electrical power distribution landscape. In our homes, we are most familiar with *single-phase* electrical power, where there are basically two power cables: a 'live' cable on which an alternating voltage is applied, given by

$$V = V_0 \sin(\omega t) \tag{12.1}$$

and a 'neutral' cable which is held at 0 V. Depending on the particular distribution network, V_0 is usually between $\sim$150 and 340 V and the frequency $\omega/2\pi$ is commonly 50 or 60 Hz. Connecting a load between the two cables completes a circuit and a current flows which will probably have a phase difference ϕ compared with the voltage, because the load may have a reactive (i.e. capacitive or inductive) as well as a resistive component, and so is given by

$$I = I_0 \sin(\omega t - \phi) \tag{12.2}$$

so that ϕ is the phase of the voltage with respect to the current.

As a result, an average real power dissipated in the load is given by

$$P = \frac{V_0 I_0}{2} \cos \phi \tag{12.3}$$

Electromagnetism for Engineers, First Edition. Andrew J. Flewitt.
© 2023 John Wiley & Sons Ltd. Published 2023 by John Wiley & Sons Ltd.
Companion website: www.wiley.com/go/flewitt/electromagnetism

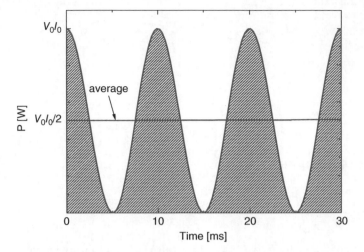

Figure 12.1 The power dissipated in a resistive load by a single-phase power supply operating at a frequency of 50 Hz.

The factor of 2 in Eq. (12.3) is a result of both the voltage and current having a sinusoidal form which means that there are two times each cycle when no current is flowing in the load and hence no power is dissipated, and the average power is half the peak power, as shown in Figure 12.1. This is exactly the situation we found when considering the power in electromagnetic waves (see Section 7.7). For this reason, we tend to work with root mean square (rms) voltages and currents, where the rms voltage or current is the peak current divided by $\sqrt{2}$. Equation (12.3) in terms of rms voltage and current becomes

$$P = V_{rms}I_{rms}\cos\phi \tag{12.4}$$

The $\cos\phi$ term is called the *power factor* and is a result of the phase difference between the voltage and current, meaning that the peak in current is out of phase with the peak in voltage and so the real power dissipated is reduced. In addition, it is helpful to define a *reactive power*

$$Q = V_{rms}I_{rms}\sin\phi \tag{12.5}$$

which is a measure of the energy stored in the reactive components of the load. By convention, as ϕ is the phase of the voltage with respect to the current, and the current in a capacitor leads the voltage by $\pi/2$, capacitive loads have a negative reactive power, while an inductor (where the current lags the voltage by $\pi/2$) has a positive reactive power. We can also define an *apparent power*

$$S = V_{rms}I_{rms} \tag{12.6}$$

which is the greatest possible power that could be dissipated for the voltage and current and is related to the real and reactive power through basic trigonometry by

$$S^2 = P^2 + Q^2 \tag{12.7}$$

The problem with single-phase electrical power is that currents tend to become very large when large real power dissipation is needed (remembering that the peak power is twice the average power) and this results in significant real power losses in the resistance of the transmission line network. Therefore, while single-phase is sufficient for electrical power demands in the home, an alternative is needed at a larger scale. *Three-phase* systems have emerged as the standard alternative. In this scheme, there are three cables, each of which carries a sinusoidal voltage, but the voltage on each cable is out of phase with respect to the other two by $2\pi/3$, as shown in Figure 12.2. The load is then connected between the three cables as three equal impedances Z, in either a *delta* formation or a *star* formation, as shown in Figure 12.3. It is possible to add a fourth *neutral* cable which is held at a potential of 0 V to the star formation. As long as the three impedances in the load are equal, the currents at the centre of the star will sum to zero and so no current flows on the neutral cable. It might therefore seem as though it would never be necessary to have a neutral cable, however, one of the attractions of the three-phase network is that a star-connected system can be used to supply to a number of houses where any individual house is provided with one phase and a neutral cable. In effect, any one house receives a simple single-phase supply and is effectively star-connected to the three-phase network. However, as long as the houses connected to each phase present a similar impedance to the network on average, then the whole system will remain *balanced* (i.e. to the three-phase network as a whole each phase will draw a similar current). The result of this is that, at any moment in time, power is always being dissipated in at least two of the three load impedances; unlike in the single-phase system there is no moment in time when there is no power dissipation. Therefore, a lower current can be used to deliver the same real power in a three-phase load.

We must now make some basic definitions of the amplitude of voltages and currents in a three-phase system. The *line voltage* V_l is the voltage between any two lines of the three-phase supply and the *phase voltage* V_{ph} is the voltage across any one phase. It is clear from Figure 12.3 that $V_l = V_{ph}$ for the delta-connected system, but that these two quantities are not equal for the star-connected system. In this case, the line voltage is greater than the phase voltage. Using the fact that the magnitude of the voltage on each phase is

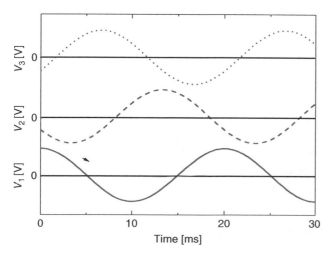

Figure 12.2 The voltages on each of the three phases of a three-phase power supply operating at a frequency of 50 Hz.

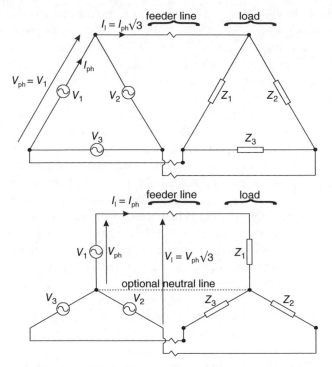

Figure 12.3 Delta (top) and star (bottom) connected three phase networks.

the same but they are out of phase with each other by $2\pi/3$, we can show that $V_l = V_{ph}\sqrt{3}$. Similarly, the *line current* I_l is the current through one line and the *phase current* I_{ph} is the current through any one phase. In this case, it is clear from Figure 12.3 that these currents are equal in the star-connected system ($I_l = I_{ph}$) but that the line current is greater than the phase current in the delta-connected system where $I_l = I_{ph}\sqrt{3}$. Taking all of this into account, it is possible to show that for either system, the total real power dissipated is given by

$$P = \sqrt{3}V_l I_l \cos\phi \qquad (12.8)$$

and the reactive power is

$$Q = \sqrt{3}V_l I_l \sin\phi \qquad (12.9)$$

Having made these basic definitions, we can now consider each of the three key parts of an electrical power network: how we generate power using synchronous machines, how we can distribute power over grids with low loss using transformers, and how we can use power in induction motors which are otherwise known as asynchronous machines.

12.2 Synchronous Machines

Large-scale electricity distribution networks, commonly called grids, have historically been based on having a relatively small number of large synchronous machines acting as generators which each provide power to the grid. A large number of users can then be connected to the grid to consume the electrical power generated. In more recent years, the situation has become more complex with the (very welcome) addition of distributed generators which harvest energy from the environment in a way that minimizes the release of carbon dioxide and other gases into the environment which can lead to global warming. These distributed generators, such as wind turbines or solar panels, increase complexity and are beyond the scope of this book. However, it is likely that synchronous machines will have at least some part to play in electrical power generation for the foreseeable future, even if just in the form of plants relying on nuclear fission or fusion.

The basic premise that underpins the operation of a large electricity power grid is that no single generator should dominate the system as a whole, and synchronous machines are designed with this in mind. At the heart of the synchronous machine is a *rotor* which, as its name suggests, is free to rotate within a stationary *stator* that surrounds it; the two are separated by a small air gap, as shown in Figure 12.4.

Our source of energy is some process which superheats steam at high pressures. Historically this would have been the burning of a fossil fuel like coal or gas, but this leads to pollution and is also costly. To avoid burning fossil fuels, many countries use nuclear fission reactors to generate heat, with perhaps one of the most notable being France which has produced a majority of its electricity in this way for many years. There is also a very significant, long-term and international research programme looking at extracting heat from nuclear fusion processes. The superheated steam is passed through highly efficient turbines which extract energy from the steam and convert it into rotational kinetic energy, and this is known as the *prime mover*. The prime mover is then connected to the rotor in the synchronous machine.

Wrapped around the rotor are a number of coils of wire which are connected to a d.c. power supply using slip rings and brushes. We know from the Ampère circuital law (see Section 3.4) that a current passing through a coil of wire will produce a magnetic field. Therefore, the rotor has a magnetic field surrounding it which will be rotating with some mechanical angular velocity ω_r. For simplicity, we will consider the situation where there is just a single coil on the rotor and so we can think of this as being like a rotating bar magnet, as shown schematically in Figure 12.5. This is called a single *pole pair*.

The stator also has coils of wire wrapped around it. For a three-phase generator we need a minimum of three coils, each one of which is connected to the grid either in a star or delta

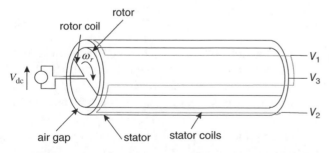

Figure 12.4 Schematic of single pole pair three-phase synchronous machine showing a d.c. voltage applied to the rotor coil and a three-phase a.c. output from the stator coils.

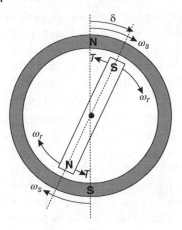

Figure 12.5 A single pole pair synchronous machine simplified in terms of north and south poles on the rotor and stator.

formation (see Figure 12.3), and these will be offset from each other physically around the stator by an angle of $2\pi/3$. Collectively, these three coils also form a single pole pair. At the most basic level, we can see that the rotating magnetic field from the rotor will cut through the three stator coils and, from the Faraday law of electromagnetic induction (see Section 3.7) this will induce an electromotive force (e.m.f.) in each of the three coils, called the *excitation*, with each being out of phase with the other two by $2\pi/3$, as required to produce the combined voltage waveforms shown in Figure 12.2.

At this level, it is clear how this basic construction of a rotor and stator combine to produce electrical power from a mechanical source of kinetic energy. However, we have overlooked a key elegance of the synchronous machine. The stator coils are connected to the three phases of the grid and so, even if we imagined completely removing the rotor assembly, there would be current flowing in each of the three coils and hence each would produce a sinusoidally varying magnetic field as a result of the Ampère circuital law. As these currents are out of phase with each other by $2\pi/3$, the coils can be arranged so that the net result is to produce a magnetic field which appears to be rotating with an angular velocity ω_s that will be equal to the angular frequency of the grid voltage. In practice, this stator magnetic field will vary sinusoidally both with angle and time around the stator, but for simplicity we can think of this as being like a single pole pair consisting of a 'north' and 'south' rotating with an angular velocity of ω_s, as shown in Figure 12.5. In fact, the system is designed so that the rotor and stator angular velocities are the same,

$$\omega_r = \omega_s \tag{12.10}$$

and rotating in the same direction. For an appropriate angle δ between the two fields (and we shall consider this in more detail shortly) the stator field interacts with the rotor field to produce a constant mechanical torque which is restraining the rotation of the rotor. The prime mover is having to do rotational work against a torque, and this is how mechanical energy is being taken from the prime mover which feels this load. The power out is then the product of this torque and the angular velocity of the synchronous machine:

$$P = T\omega_s \tag{12.11}$$

The torque will vary with the physical angle of the rotor field with respect to the stator field, which is known as the *load angle*. If the load angle is zero, then the two fields align and no torque is exerted by the stator on the rotor. As the load angle increases, the restraining torque will also increase as

$$T = T_0 \sin \delta \tag{12.12}$$

up to a maximum T_0 for a given excitation.

The consequence of this is that the magnetic field produced by the rest of the grid on the stator ensures that the synchronous machine is in phase electrically with the rest of the grid: this is where the synchronous machine gets its name from. Furthermore, at any moment in time, the total power output by all of the generators that are connected to the grid must match the power demanded by the loads. Therefore, any one synchronous machine will sit at the appropriate load angle to ensure that it is providing sufficient power output to match the demand. If the power demand rises, then the load angle of all of the connected generators will increase to raise the torque and hence the power output. Clearly, we do not want to get to a point where the maximum torque T_0 is exceeded, otherwise the stator field will no longer be able to restrain the rotor. Therefore, if the load angle gets too large, additional generation capacity must be brought online to reduce the demand on the generators, either by increasing the excitation of existing generators or by connecting more generators. If demand falls, then the load angle will drop once more, and generation capacity may be safely reduced. It is even possible for a generator to absorb excess energy produced elsewhere in the system by allowing the load angle to become negative (the stator field is dragging round the rotor).

This is best understood quantitatively by considering the equivalent circuit for the three-phase synchronous machine, one phase of which is shown in Figure 12.6(a). We can consider the excitation on one phase as being an a.c. voltage source $\widetilde{E_{ph}}$ (where the tilde represents a phasor as opposed to a magnitude). The coils of the rotor and stator clearly have a significant inductance, and so can be modelled as a series reactance X_s. Therefore, for a given phase current $\widetilde{I_{ph}}$, applying the Kirchhoff voltage law around the loop shown in Figure 12.6 gives

$$\widetilde{E_{ph}} = \widetilde{V_{ph}} + jX_s\widetilde{I_{ph}} \tag{12.13}$$

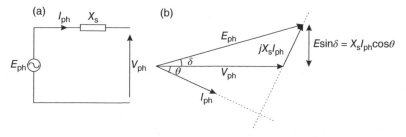

Figure 12.6 (a) Equivalent circuit and (b) Phasor diagram for one phase of a three-phase synchronous machine, showing how the load angle can be related to the current drawn by the grid.

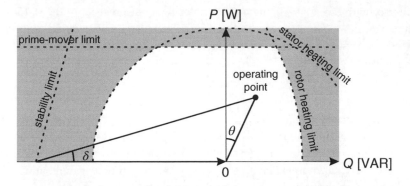

Figure 12.7 Operating chart for the same state of the three-phase synchronous machine from Figure 12.6 achieved by multiplying axes by $3V_{ph}/X_s$. The unshaded area represents the region in which the synchronous machine may be safely operated.

We can represent Eq. (12.13) on a phasor diagram as shown in Figure 12.6(b). Here we will assume that the grid is presenting an inductive load, so the phase current is lagging the phase voltage. This is normally the case on the grid to ensure stability. In the phasor diagram we choose a moment in time when the voltage phasor is pointing along the positive real axis. The current phasor will therefore be pointing into the positive real/negative imaginary quadrant. According to Eq. (12.13), we must add a $jX_s\widetilde{I_{ph}}$ phasor to the phase voltage, and this will point in a direction at right angles to $\widetilde{I_{ph}}$. The excitation phasor is then the sum of these two. Trigonometry then allows the power factor and the load angle to be related to each other by equating the imaginary component of the excitation phasor to give

$$E_{ph} \sin \delta = X_s I_{ph} \cos \theta \tag{12.14}$$

This equation formalizes quantitatively how an increase in the current drawn by the grid must be matched by an increase in the load angle, while increasing the excitation allows the load angle to be reduced.

We can also convert the phasor diagram of Figure 12.6(b) into a power operating chart by multiplying both axes by $3V_{ph}/X_s$ and moving the origin of the graph to the base of the $jX_s\widetilde{I_{ph}}$ phasor, as shown in Figure 12.7. This results in the vertical axis representing real power and the horizontal axis representing reactive power. Key system limitations can then be plotted on this chart to find the regions in which the system can be safely operated. Most notable among these limitations, each of which are shown in Figure 12.7, are:

- the *prime-mover limit* which is the maximum power output that the prime mover can support;
- the *stator heating limit* which effectively limits the maximum phase current;
- the *rotor heating limit* which defines the maximum excitation of the system; and
- the *stability limit* which is the maximum load angle that is considered tolerable.

It is also clear from the operating chart that, by controlling the excitation, the synchronous machine can generate real power output while also producing either positive or negative reactive power. This allows the generators to be used to help define the overall power factor of the network, ensuring a net positive reactive power for the system and hence stability.

Finally, although we have only considered a synchronous machine with a single pole pair, it is possible to increase the number of coils on the stator to create more pole pairs. This has the effect of reducing the angular velocity of the stator field. For a grid angular frequency of ω, the stator angular frequency will be

$$\omega_s = \frac{\omega}{p} \tag{12.15}$$

where p is the number of pole pairs. In order for a non-zero torque to act on the rotor, the number of pole pairs on the rotor must also be increased to be equal to p. This allows some flexibility in the design of the synchronous machine so that the mechanical rotation speed of the rotor is such that the torque produced by the prime mover is large (a diesel generator, for example, produces maximum torque at a much lower speed than a steam turbine).

To summarize, in the synchronous machine we have two rotating magnetic fields: one created by the a.c. current from the grid on the stator coils and one created by the d.c. current on the rotor coils which is made to rotate by the prime mover. The two fields have the same angular velocity with a load angle between them, and the prime mover does work against a force as the load angle means that the rotor is being restrained by the stator field. By induction, this work is converted into electrical power to the grid.

12.3 Transformers

A key advantage of electrical power is that we are able to spatially separate where we generate electrical power, for example using synchronous machines, from where we use electrical power in homes, factories, trains and elsewhere. Therefore, we need a distribution network, commonly called a grid, to connect generation plants to end users. This inevitably requires the use of long-distance cables which will have a finite resistance. As the power dissipated on a cable is proportional to I_l^2, it is essential to minimize the line current. This can be achieved by using high voltages for the long-distance part of the grid, as conservation of energy means that if the voltage is increased by a factor of 10, then the current is decreased by a similar factor. For example, the UK's long-distance grid operates at either 275 or 400 kV and is gradually stepped down through more local parts of the network to 132, 33 and 11 kV before being reduced to 230 V (rms) for use by most consumers.

These step changes in voltage are achieved using *transformers*. Although there are several specific designs of transformer, the basic principle is that one phase of the three-phase supply is connected to a coil of wire, called the *primary coil*, with N_1 turns wrapped around an arm of a ring of soft ferromagnetic material called the *core* of the transformer. This creates a solenoid and so the alternating current passing through the primary coil creates a sinusoidal magnetic flux in the ferromagnetic core which itself forms a magnetic circuit (see Section 4.7). This primary coil has a magnetomotive force of N_1I_1 where I_1 is the rms current in the primary coil. A second coil with N_2 turns, known as the *secondary coil*, is also wrapped around the core, as illustrated in Figure 12.8. The sinusoidal magnetic flux passing through this secondary coil results in an e.m.f. being induced which produces a current I_2 in a load that is connected to the output of the secondary coil. The magnetomotive force of the secondary coil will be the same as for the primary coil, and hence

$$N_1I_1 = N_2I_2 \tag{12.16}$$

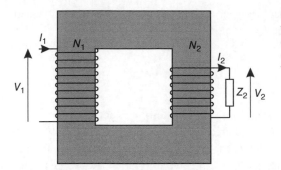

Figure 12.8 Schematic of a simple transformer with a primary and secondary coil wrapped around a soft ferromagnetic core with a load impedance connected to the secondary coil.

As power must be conserved, it must be the case that

$$V_1 I_1 = V_2 I_2 \tag{12.17}$$

and so substitution into Eq. (12.16) gives

$$\frac{V_2}{V_1} = \frac{N_2}{N_1} \tag{12.18}$$

In other words, the ratio of the number of turns on the primary and secondary coil can be used to define the ratio by which the voltage and current are increased or decreased. Therefore, by having such a transformer on each phase of a three-phase supply, the voltage can be modulated for different parts of the grid.

When performing analysis of circuits, the presence of a transformer can be inconvenient. However, we can simplify things using the principle of *referral*. Let us imagine that a load impedance Z_2 is connected to the secondary side of a transformer as shown in Figure 12.8. This will cause a current I_2 in the secondary coil and hence a current I_1 in the primary coil. The apparent input impedance of the primary side of the transformer, which we will call Z_2', is given by

$$Z_2' = \frac{V_1}{I_1} \tag{12.19}$$

However, we can substitute for V_1 and I_1 using Eqs. (12.16) and (12.17) to give

$$Z_2' = \frac{V_2}{I_2} \left(\frac{N_1}{N_2} \right)^2 \tag{12.20}$$

and V_2/I_2 is just the impedance connected to the secondary side, so

$$Z_2' = Z_2 \left(\frac{N_1}{N_2} \right)^2 \tag{12.21}$$

Therefore, we can consider the transformer and its load impedance Z_2 to be equivalent to a simple impedance Z_2' which we call the referred load impedance.

In practice, a transformer will not be ideal, and so the equivalent circuit of Figure 12.9 is employed. It consists of both series resistances and reactances on each of the primary (R_1 and X_1) and secondary (R_2 and X_2) sides of the transformer and a single set of parallel resistance R_0 and reactance X_0 on the primary side which sit around an ideal transformer. R_1 and R_2 represent the resistance of the primary and secondary coils, respectively. X_1 and X_2 represent the imperfect coupling of the magnetic flux between the two coils which appears

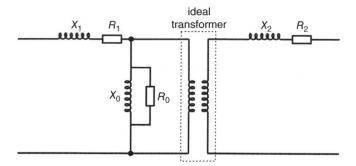

Figure 12.9 Equivalent circuit for a non-ideal transformer.

as a self-inductance for each coil. R_0 models the power losses within the transformer's core, which are usually a result of both the finite hysteresis of the ferromagnetic material used in the core (see Figure 4.5) and the formation of eddy currents in the core. These are circulations of electrons around the magnetic flux lines and are suppressed in practice by making the core out of thin sheets of ferromagnetic material with thin insulating layers between them that largely prevent the circulating currents from forming. Finally, X_0 represents the finite reactance of the core. In practice, R_2 and X_2 can also be referred to the primary side of the transformer along with the load to simplify circuit analysis.

12.4 Asynchronous Machines

So far in this chapter, we have considered how we generate electrical power using synchronous machines and why we do this in such a way as to create a three-phase supply for large-scale electricity distribution networks (grids). We have also seen how transformers can be used to modulate the amplitude of the voltage and current to ensure power loss on the grid is minimized. Finally, we now consider a device that is connected to the grid as a load. In reality there are a vast number of such devices, but in this book on electromagnetism, it seems appropriate to focus on just one: the asynchronous machine, which is more commonly called an *induction motor*. Not only do induction motors rely on electromagnetism to operate, but they are also a commonly used motor driven by an a.c. voltage.

In many respects, the induction motor is very similar to the synchronous machine in its construction, as shown in Figure 12.10. We have a central rotor that is wrapped in coils of wire and which is free to rotate. This is surrounded by the stationary stator which is also wrapped in coils of wire.

Just as with the synchronous machine (see Section 12.2), the stator coils are connected to the three-phase grid to create a rotating magnetic field. In its simplest construction, there could be just one coil connected to each phase of the supply with the three coils at an angle of $2\pi/3$ with respect to each other. This will create a rotating magnetic field on the stator ω_s whose angular velocity matches that of the grid's angular frequency. However, once again in a similar fashion to the synchronous machine, we can create more pole pairs by adding coils to the stator, and so if we have p pole pairs on the stator, then the angular velocity of

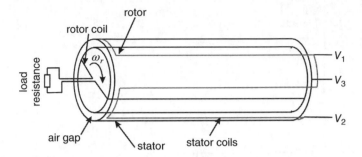

Figure 12.10 Schematic diagram of the induction motor (asynchronous machine).

the stator field will be

$$\omega_s = \frac{\omega}{p} \tag{12.22}$$

The coils on the rotor must have the same number of pole pairs as the stator and each phase of the coil is connected to a load resistance R_2 (which may be zero) using slip rings, but there is no other direct electrical connection to these coils. This is the key difference between the asynchronous and synchronous machine. Let us imagine that the rotor is initially stationary. The rotating stator field will cut through the rotor coils and induce an e.m.f. This, in turn, will produce a sinusoidal current in the coils and hence a rotor magnetic field which will interact with stator field to produce a torque on the rotor and therefore a mechanical rotation.

The key to understanding the induction motor lies in appreciating exactly how the rotor and stator fields interact with each other. This is best done by imagining that the rotor is mechanically rotating at an angular velocity ω_r where a positive value indicates rotation in the same direction as the stator field. At this point, it is helpful to define a quantity called *slip*, which is the fractional difference between the mechanical angular velocity of the rotor and the angular velocity of the stator field:

$$s = \frac{\omega_s - \omega_r}{\omega_s} \tag{12.23}$$

If the rotor is not moving, then the slip is 1, but if the rotor is rotating at the same speed as the stator field then the slip is zero. We can also rearrange Eq. (12.23) to express the angular velocity of the rotor in terms of slip as

$$\omega_r = (1 - s)\omega_s \tag{12.24}$$

Let us now consider how the stator field appears to the rotor. In the rotor's frame of reference, the stator field will appear to be rotating with an angular velocity of $\omega_s - \omega_r$, which can be rewritten using Eq. (12.23) in terms of slip as $s\omega_s$. Substituting in terms for the angular frequency of the three-phase supply using Eq. (12.22) gives an angular velocity of $s\omega/p$ for the stator field in the frame of reference of the rotor and therefore this will also be the angular velocity of the field induced in the rotor by the stator field in the same frame of reference. If we now return to the static frame of reference of the stator, the rotor field will appear to have an angular velocity of $s\omega/p + \omega_r$ which is just equal to ω_s. Therefore, the rotor and stator fields have the same angular velocity and a constant torque acts on the rotor at a particular speed of mechanical rotation.

From an electrical circuit perspective, the induction motor is actually very similar to the transformer, and so a very similar equivalent circuit is used for both. To begin, let us imagine that we use the equivalent circuit for the transformer in Figure 12.9 in the situation where the rotor has been 'locked' and is not allowed to move so that the slip $s = 1$. The only change is that we short-circuit the coils on the rotor, and so we must imagine short-circuiting the secondary side of the transformer equivalent circuit. The stator is like the primary side of the transformer with loss terms of R_1 and X_1 representing the resistance of the coils and leakage reactance, respectively. The parallel terms on the primary side R_0 and X_0 represent the real power losses in the core of the induction motor and the reactance associated with creating the magnetic field between the stator and rotor. We then have the ideal transformer which represents the ideal flux linkage between the rotor field and the stator field. In the case of the transformer, the ratio of the number of physical turns defines the voltage ratio between the primary and secondary side as a soft ferromagnetic core links the two. The situation is more complex in the induction motor, as there is an air gap between the stator and rotor and a cylindrical geometry. Therefore, we define a simple effective turns ratio k which defines the ratio of the voltage on the stator side of the ideal transformer E_1 relative to the induced e.m.f. on the rotor E_2 for the situation where the rotor has an open circuit on the coils (which of course we do not do as we need a current to flow to generate torque). Finally, on the rotor side we define R_2 as the resistance of the rotor coils and any additional load resistance that is connected to the coils, and X_2 as the leakage reactance of the rotor.

If we now allow the rotor to move, then the slip $s < 1$. This proportionally reduces the e.m.f. induced in the rotor coils which now becomes sE_2 (i.e. the e.m.f. reduces to zero when the rotor angular velocity is the same as the stator field angular velocity, at which point $s = 0$). Also, as X_2 is an inductance, its magnitude is proportional to frequency (as the impedance of an inductor is $j\omega L$). We have already seen that the angular velocity of the field induced in the rotor is proportional to slip, and therefore X_2 in the equivalent circuit for the induction motor becomes sX_2. We can now use the Kirchhoff voltage law to sum voltages around a loop on the rotor side of the equivalent circuit to give

$$sE_2 = I_2(R_2 + jsX_2) \tag{12.25}$$

where I_2 is the phase current in the rotor coils. If we divide both sides of this equation by s then we have

$$E_2 = I_2 \left(\frac{R_2}{s} + jX_2 \right) \tag{12.26}$$

and we can use this as the basis for creating an equivalent circuit as shown in Figure 12.11 that is specific for the induction motor.

This final step of dividing all the terms on the rotor side of the equivalent circuit by a factor of s is particularly important. It means that we have an ideal transformer in the middle of the circuit defined by a simple constant k. We can now employ the idea of referral that we saw for the transformer in Eq. (12.21). This states that we can replace a real circuit containing a transformer with an equivalent circuit where the transformer is removed. The factor k^2 allows us to refer all of the impedances on the rotor side of the induction motor to the stator

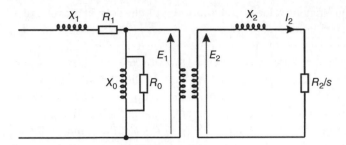

Figure 12.11 The equivalent circuit of the induction motor. In the centre is an ideal transformer with an effective turns ratio of k.

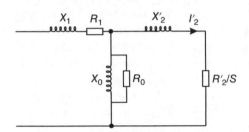

Figure 12.12 The equivalent circuit of the induction motor where the terms on the rotor side have been referred to the stator.

side, and the resulting equivalent circuit is shown in Figure 12.12. This circuit can be used to analyse power dissipation in the induction motor – both power dissipated as losses and power dissipated as mechanical torque.

The power dissipated as mechanical torque is actually represented within the R'_2/s term. This is made up of two components. The actual referred resistance of the rotor coil and any added load resistance is actually still R'_2, which is less than R'_2/s as $s \le 1$. The difference between the two is

$$R'_2 - \frac{R'_2}{s} = R'_2 \left(\frac{1-s}{s} \right) \tag{12.27}$$

It is this resistance that is modelling the real power out that is converted into electromechanical torque T. We can therefore equate the power dissipated in this resistance with the mechanical power, which is $T\omega_r$, to give

$$T = \frac{3I'^2_2}{\omega_r} R'_2 \left(\frac{1-s}{s} \right) \tag{12.28}$$

where I'_2 is the referred current in the rotor and the factor of 3 is because we are using a three-phase supply. Equation (12.28) can be further simplified by substituting Eq. (12.24):

$$T = \frac{3I'^2_2 R'_2}{s\omega_s} \tag{12.29}$$

Based on Eq. (12.29), Figure 12.13 shows the typical form of a torque curve for an induction motor as a function of rotor angular velocity. There are four general regions of operation. Induction motors should generally be used at rotor angular velocities where the slip is between zero and where the slip gives the maximum torque. In this region, if the load on the rotor is increased, then the rotor slows down but the torque increases. As long as the load does not exceed the maximum torque that can be supplied, the rotor will

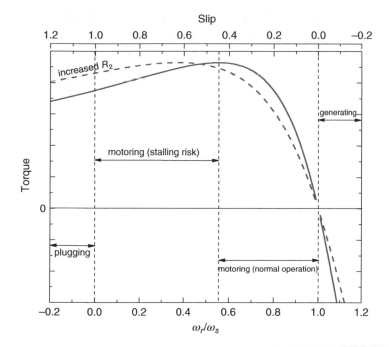

Figure 12.13 The torque as a function of the ratio of the rotor angular velocity to the stator angular velocity. The equivalent slip is shown on the top axis for information.

be able to keep rotating. If this maximum torque is exceeded, then the induction motor will move into the region between where the slip gives maximum torque and where the slip is 1. In this region, the motor will stall if the load requires more torque than can be delivered as the torque actually decreases with reducing angular velocity of the rotor rather than increasing. Therefore, we want to try and avoid getting too close to this condition in real life. Fortunately, we can adjust the shape of the torque curve by adding load resistance to R_2, as shown in Figure 12.13. This moves the slip at which torque is maximized to lower angular velocities. Having a controlled variable resistance connected to the rotor coils therefore allows us to ensure that we are always in the first region and have a large torque available. This is a key advantage of electric motors as they do not need gears. In contrast, combustion engines, for example, produce maximum torque over a fixed and relatively narrow range of angular velocities, and therefore gears have to be used to ensure that torque is maximized for a given output rotation speed, which complicates the transmission system.

In both of these first two regions (for $0 \geq s \geq 1$), the induction motor is said to be *motoring*. We could, of course, attach another source of mechanical torque to the rotor. If we use such a source to deliberately drive the rotor in the opposite direction to the stator field ($s > 1$) the motor is said to be *plugging*. Of more practical use, however, is when we drive the rotor in the same direction as the stator field but at a higher angular velocity. In this *generating* condition, the asynchronous machine is being used to produce electrical power which will be output on the stator.

To summarize, for the asynchronous machine, the stator coils are connected to a three-phase supply which produces a rotating magnetic field, as was the case for the synchronous machine as well. However, the rotor coils are connected to a short circuit or a simple load resistance. The stator field induces a current in the rotor coils which in turn produces a rotating rotor field that always has the same angular velocity as the stator field. This ensures that a constant torque is produced on the rotor which is therefore driven round.

Epilogue

Although my expectation is that most readers of this book will just read individual chapters at a time, there may be a few readers who decided to read the book in its entirety. For such readers, I hope the experience has been an enjoyable journey. Our starting point was to consider some very fundamental ideas about the origin and nature of electric and magnetic fields and how they interact with matter. There are a few basic assumptions that, as engineers, we made at this point. These avoided the need to get too side-tracked early on by some detailed physics. However, having made these assumptions, we were able to set up some very basic equations that quantify how electric and magnetic fields behave. At this point, the Maxwell equations allowed us to express this behaviour in a very elegant mathematical way. This then let us study a wide diversity of applications.

What I hope an engineer will take away from this is the power of these fundamental equations. We have only been able to look at a relatively small (but hopefully important) subset of applications of electromagnetism. In each case, there is a common pattern: firstly, to think about the system that we are dealing with and how this impacts the form of the Maxwell equations, the charges and currents that are present and any boundary conditions that might need to be applied; secondly, to then solve the equations that we have generated for the given situation; and finally, to think carefully about how to interpret what the result tells us about how the system is behaving. In this way, we can hope to engineer the system from a position of understanding.

We face many societal challenges in the years ahead which engineers must help to address. Arguably the most significant is around climate change. How do we live well but with minimal environmental impact? We will need not only to be able to generate electrical power more sustainably using devices like solar cells and wind turbines, but also to use energy more efficiently, and this means tackling the high energy demands of both processing and communicating data.

Electromagnetism will be at the centre of our ability to engineer this better future.

Electromagnetism for Engineers, First Edition. Andrew J. Flewitt.
© 2023 John Wiley & Sons Ltd. Published 2023 by John Wiley & Sons Ltd.
Companion website: www.wiley.com/go/flewitt/electromagnetism

Index

Electromagnetism for Engineers, First Edition. Andrew J. Flewitt.
© 2023 John Wiley & Sons Ltd. Published 2023 by John Wiley & Sons Ltd.
Companion website: www.wiley.com/go/flewitt/electromagnetism